NEW YORK STATE GRADE 7

MATH TEST

Amy Stahl, M.S. Ed.
7th Grade Math Teacher
Victor Junior High
Ontario County

ACKNOWLEDGMENTS

I cannot even begin to thank the people who have helped me to become the person and teacher I am today. Special thanks to

- Terry Neal, Mathematics Professor at Genesee Community College, to whom this book is dedicated in memory of. Thank you, because you always knew the kind of teacher I would be, even before I knew it myself.
- all my students and their families.
- my colleagues at Victor Junior High and former colleagues at TC Armstrong Middle School and Fairport High School.
- Annemarie McNamara at Barron's.
- my loving and supportive husband, family, and friends. Without you, this would not have been possible.

All inquiries should be addressed to:
Barron's Educational Series, Inc.
250 Wireless Blvd.
Hauppauge, NY 11788
www.barronseduc.com

ISBN-13: 978-0-7641-4070-9
ISBN-10: 0-7641-4070-1

Library of Congress Catalog Card No. 2008014244

Library of Congress Cataloging-in-Publication Data

Stahl, Amy.
New York State grade 7 math test / Amy Stahl.
p. cm.
Includes index.
ISBN-13: 978-0-7641-4070-9
ISBN-10: 0-7641-4070-1
1. Mathematics—Examinations, questions, etc.—Study guides. 2. Mathematics—Study and teaching (Elementary)—New York (State) 3. Seventh grade (Education)—New York (State) 4. Examinations—New York (State)—Study guides. I. Title.

QA43.S774 2008
510.76—dc22

2008014244

Printed in the United States of America

9 8 7 6 5 4 3 2 1

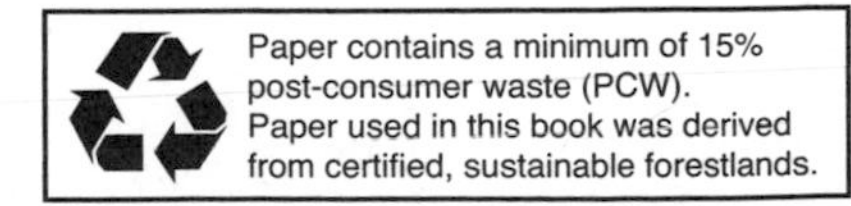

Paper contains a minimum of 15% post-consumer waste (PCW).
Paper used in this book was derived from certified, sustainable forestlands.

CONTENTS

INTRODUCTION

This review book will help you get ready for the NYS Grade 7 Math Test. To meet the requirements of the No Child Left Behind (NCLB) Act, since 2006, seventh graders are required to take a state math test that is administered annually. The test is usually given in March, depending on state guidelines and your school's district calendar. The test is to serve as a measure of progress and provides information about student preparedness for study at the next level. The test will help identify students who may be in need of additional academic assistance in math. The test alone does not determine promotion or retention.

Everyone in New York State will take the NYS Grade 7 Math Test. The test is timed and consists of two sections. The table below outlines the sections of the test:

Day 1 Book 1 No calculator	30 multiple choice questions Students provided with a scantron	55 minutes, plus 10 minutes prep time
Day 2 Book 2 Scientific calculator permitted Use of formula reference sheet provided by the state	4 short-response questions 4 extended-response questions Students write directly in test booklet	55 minutes, plus 10 minutes prep time

The tests are scored by a group of teachers using guidelines provided by the state.

So that all students are graded equally, the New York State Education Department distributes rubrics and scored sample answers to each school district, to be used by the teachers who grade the exams. Scores on the assessment range from 1 to 4. A score of 4 means the student's performance exceeds the state standards. A score of 3 means the student's performance meets the state standards. A score of 1 or 2 means the student's performance is below state standards. If a student receives a score of 1 or 2, the student will then receive additional instruction or academic intervention services to meet the state standards. Check with your individual school to see how these services will be administered.

At the end of each section there are practice questions similar to those on the New York state test. To further help you to prepare for the state test, there are two full practice exams at the end of the book, followed by answers and explanations to all of the quiz and test questions.

As you are working through the book, here are some strategies that will help you to be successful:

- Read each question carefully.
- Underline key words, numbers, and phrases.
- Show work for each problem.
- Cross off multiple-choice answers that do not make sense.
- After you solve the problem, reread the problem again to make sure you have answered all the parts of the question.

Also remember to use strategies that you have been taught in math class:

- Guess and check
- Draw a picture or a diagram
- Make a list
- Solve a simpler related problem
- Work backwards
- Look for a formula
- Think aloud

TO THE STUDENT

It is important to remember that studying for math is different than studying for any other subject. Here are some tips you can use while taking the test.

- Listen to all of the instructions and directions.
- Do the problems you know you can do first! This will leave you more time to do the problems that you find more difficult.
- Keep track of the time.
- Show all of your work.
- Read the questions carefully.
- For multiple-choice questions, cross off answers that you know do not make sense.
- If you are stuck, try to use the answers to work backwards.
- Check that your answers make sense.
- Watch for careless mistakes (forgot a decimal, wrote the wrong sign, etc.).
- Complete all sections of the problem, even if Part A is confusing, complete Part B and Part C by doing the best that you can.
- If you finish early, go back and check yourself. Redo the problems if there is time.

Most importantly, do the best that you can. We hope that you find this book a valuable resource as you prepare for the New York state test.

Good luck!

Chapter 1

NUMBER SENSE AND OPERATIONS

OPERATIONS WITH INTEGERS

If you want to learn to drive a car, you must first learn the rules of the road. To be successful with positive and negative numbers, you must first learn the rules of integers!

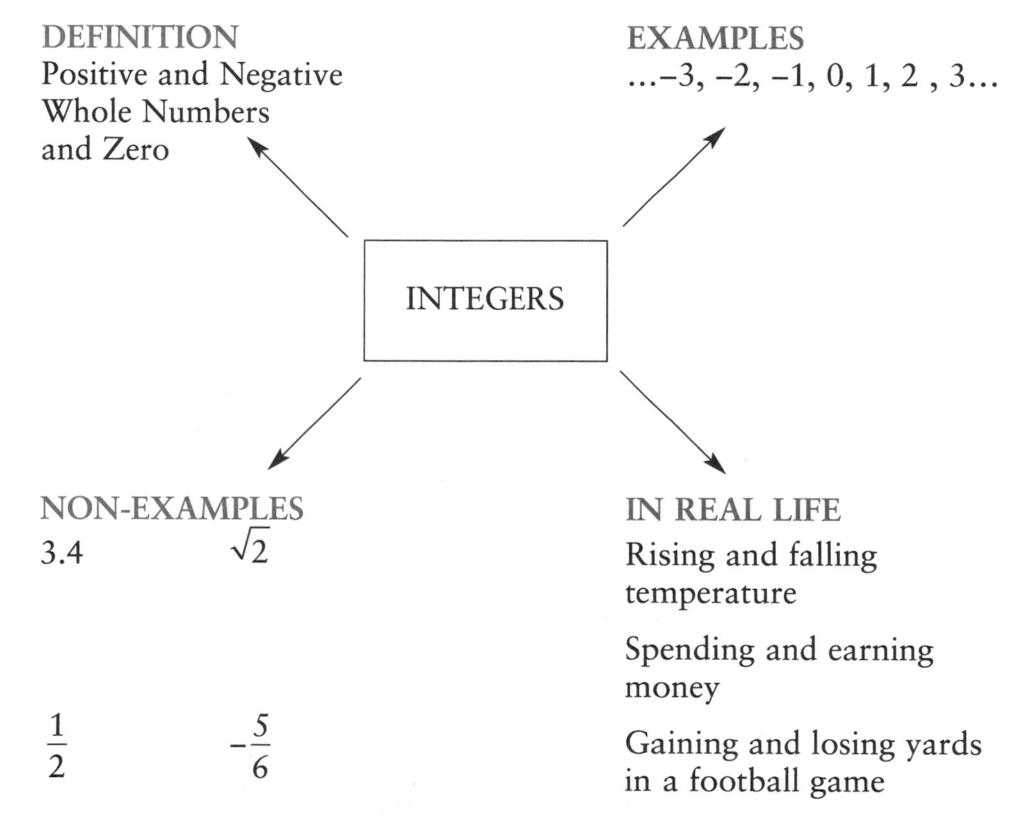

Positive means direction. Positive means move to the right or up from zero.
Negative also means direction. Negative means move to the left or down from zero.

INTEGER ADDITION RULES

RULE 1: Signs same, add, keep same sign.
RULE 2: Signs different, subtract, take the sign of the number with the largest absolute value.

Examples

1. $-3 + -5$	Sign same, add, keep same sign	-8
2. $6 + 10$	Sign same, add, keep same sign	16
3. $5 + -8$	Signs different, subtract	$8 - 5 = 3$
	Since $\lvert -8 \rvert = 8$ is the larger absolute value, the answer is negative	-3
4. $12 + -6$	Sign different, subtract	$12 - 6 = 6$
	Since 12 is the larger absolute value, the answer is positive	

INTEGER SUBTRACTION RULE

RULE: Keep first number, change subtraction to addition, write opposite of second number (KAO).
Then follow the rules for addition.

Examples

	K	A	O		
1. $4 - 9 =$	4	$+$	(-9)	$=$	-5
2. $-2 - 8 =$	-2	$+$	-8	$=$	-10
3. $4 - (-3) =$	4	$+$	3	$=$	7
4. $-3 - (-7) =$	-3	$+$	7	$=$	4

INTEGER RULES FOR MULTIPLYING AND DIVIDING

(for two numbers only!)

RULE 1: Signs same, answer positive.
RULE 2: Signs different, answer negative.

A pizza story to help you remember the multiply and divide rules:

+	+	+	"I like pizza, you like pizza, we agree"
+	–	–	"I like pizza, you don't like pizza, we disagree"
–	–	+	"I don't like pizza, you don't like pizza, we agree"

Examples

1. $(-4)(3) =$ -12 Signs different, answer negative or "I don't like pizza, you like pizza, we disagree"

2. $(-5)(-5) =$ 25 Signs same, answer positive or "I don't like pizza, you don't like pizza, we agree"

3. $2(8) =$ 16 Signs same, answer positive or "I like pizza, you like pizza, we agree"

4. $\frac{-12}{6} =$ -2 Signs different, answer negative or "I don't like pizza, you like pizza, we disagree"

5. $\frac{-24}{-8} =$ -3 Signs same, answer positive or "I don't like pizza, you don't like pizza, we agree"

Remember to use the rule for two numbers only, and then do it again.

TEST YOUR SKILLS (For answers, see page 237.)

1. What is the sum of –25 and –36?

   A. 61
   B. –61
   C. 11
   D. –11

2. Find the value of –23 + 7.

   A. –30
   B. 30
   C. 16
   D. –16

3. Find the sum of –6 and 15.

   A. 9
   B. –9
   C. 21
   D. –21

4. Find the value of 48 – (–24).

   A. 24
   B. –24
   C. 72
   D. –72

5. Find the value of $-8(9)$.

   A. -17

   B. 17

   C. -72

   D. 72

6. Find the product of -4 and -11.

   A. 44

   B. -44

   C. -15

   D. 15

7. Find the value of the expression $(-2)(3)(-4)$.

   A. 9

   B. -9

   C. 24

   D. -24

8. $\frac{-36}{-4} =$ _____

   A. -9

   B. 9

   C. -32

   D. 32

9. What is the quotient of -72 and 9?

   A. -8

   B. 8

   C. -63

   D. 63

10. Water cools at a rate of 3 degrees per hour. At that rate, how long will it take to reach –12°F if the water temperature is currently 6 degrees?

A. 3 hours

B. 4 hours

C. 5 hours

D. 6 hours

11. Which is *not* a true statement?

A. $-12 + 3 = -9$

B. $-12 - 3 = -15$

C. $-12(3) = -36$

D. $\frac{-12}{-3} = -4$

Short Response

12. You have been watching the stock market on the news every night. Your math teacher has asked you to watch the stock for Apple and report it to her after a week. During the course of a week, here is what you found: on Monday the stock started at (–8) points; on Tuesday the stock went up 3 points; on Wednesday the stock dropped 10 points; on Thursday the stock dropped another 4 points; and on Friday the stock rose 5 points.

Part A.

What is the value of the Apple stock by the end of the week? **Show your work.**

Answer –6

Part B.

How many points would the stock need in order to reach zero? **Show your work.**

Answer 6

EXPONENTS

An expression like $2 \cdot 2 \cdot 2 \cdot 2$ can be written using exponents as 2^4.

The 2 is called the base. A base is the "bottom" number. It is the number that is multiplied.

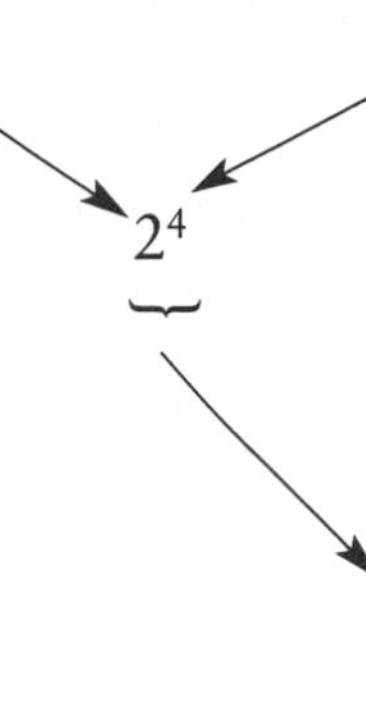

The 4 is called the exponent. The exponent tells you how many times you write out the base using multiplication.

The number that is expressed using an exponent is called a power.

Power	How to say it	How to write it
3^1	3 to the first power	3
3^2	3 to the second power or 3 squared	$3 \cdot 3$
3^3	3 to the third power or 3 cubed	$3 \cdot 3 \cdot 3$
3^4	3 to the fourth power	$3 \cdot 3 \cdot 3 \cdot 3$

LAWS OF EXPONENTS

MULTIPLYING

Examples

$2^2 \cdot 2^3 = \quad 2 \cdot 2 \cdot 2 \cdot 2 \cdot 2 = \quad 2^5$

$3^4 \cdot 3^7 = \quad 3 \cdot 3 \cdot 3 \cdot 3 \cdot 3 \cdot 3 \cdot 3 \cdot 3 \cdot 3 \cdot 3 \cdot 3 = \quad 3^{11}$

RULE: When multiplying, if the bases are the same, *add* the exponents.

DIVIDING

Examples

$$\frac{4^5}{4^3} = \frac{4\cdot4\cdot4\cdot4\cdot4}{4\cdot4\cdot4} = 4^2$$

$$\frac{x^7}{x^4} = \frac{x\cdot x\cdot x\cdot x\cdot x\cdot x\cdot x}{x\cdot x\cdot x\cdot x} = x^3$$

RULE: When dividing, if the bases are the same, subtract the exponents.

POWER TO A POWER

Examples

$(2^3)^3 = 2^3 \cdot 2^3 \cdot 2^3 = 2^9$

RULE 1
If you write the problem out using multiplication, use the multiply rule and add the exponents.

$(2^3)^3 = 2^9$

RULE 2
If you use the power to a power rule, multiply the exponents.

TEST YOUR SKILLS (For answers, see page 238.)

1. Find the value of $2^3 \cdot 2^4$.

   A. 2^5
   B. 2^7
   C. 2^{12}
   D. 4^{12}

2. $3^2 \cdot 3^3 =$

A. 45

B. 54

C. 81

D. 243

3. $(2^3)^2 =$

A. 2^5

B. 2^6

C. 2^{12}

D. 2^{16}

4. Find the value of $\frac{4^3}{4^1}$.

A. 4

B. 8

C. 16

D. 42

5. Evaluate: $\frac{5^4}{5^2}$

A. 5

B. 10

C. 25

D. 34

6. $(4^5)^2 =$ _____

A. 4^7

B. 4^{10}

C. 8^{10}

D. 4^{16}

7. Compare: 3^2 _____ 2^3

A. $>$

B. $<$

C. $\geq$

D. $=$

Short Response

8. You are learning about cell division in science class. A certain type of bacteria will split into two cells every 10 minutes.

Part A.

2 → 10 mn.

If you are starting with one cell, how many cells will there be in 30 minutes? Show your work.

Answer ___8___

Part B.

Use what you know about exponents to explain why your answer in Part A is correct.

Since two cells will split every 10 mn, the first time it will split two, then four, and lastly 8, 2^3.

ORDER OF OPERATIONS WITH ABSOLUTE VALUE, INTEGERS, AND EXPONENTS

How would you solve $4 + 6 \cdot 9$? Do you add first or multiply? In order to get the correct answer, you need to follow the order of operations.

The phrase "Please Excuse My Dear Aunt Sally" will help you remember the order of the rules.

P	parentheses	→	Do the operation inside the parentheses first.
E	exponents	→	Simplify exponents.
MD	multiply/divide	→	Multiply and divide in order from left to right.
AS	add/subtract	→	Add and subtract in order from left to right.

Examples

1. $15 + (20 - 3) + 2^3$ — Parentheses? Yes—do the subtraction in the parentheses.

   $15 + 17 + 2^3$ — Exponents? Yes—simplify the exponent.

   $15 + 17 + 8$ — Now add from left to right. Add 15 and 17.

   $32 + 8$ — Add 32 and 8.

   40

2. $8 - 3(2) + 7$ — Parentheses? No—the parentheses in this problem represents multiplication. There is no operation *inside* the parentheses. So first multiply 3(2).

   $8 - 6 + 7$ — Now subtract and then add, from left to right. Subtract 8 minus 6.

   $2 + 7$ — Add 2 + 7.

   9

3. $5(2) + 18 \div 3$ — Since there is multiplication and division in the same problem, work left to right. Multiply 5(2).

$10 + 18 \div 3$ — Now divide $18 \div 3$.

$10 + 6$ — Add $10 + 6$.

16

4. $4(3^3 - 2^2)$ — Evaluate the exponents in the parentheses.

$4(27 - 4)$ — Do the subtraction inside the parentheses.

$4(23)$ — Multiply 4(23).

92

5. $5^2 + \frac{7(10)}{2+3}$ — Simplify the numerator and the denominator of the fraction.

$5^2 + \frac{70}{5}$ — Evaluate the exponent.

$25 + \frac{70}{5}$ — Divide.

$25 + 14$ — Add.

39

6. $|-10 + 7|$ — Simplify inside the absolute value bars.

$|-3|$ — Take the absolute value of 3.

3

TEST YOUR SKILLS (For answers, see page 239.)

1. Simplify the expression: $|-5| + 9 \times 6 \div 2$

A. 29.5

B. 32

C. 42

D. 64

2. Simplify: $4(8 - 3) + 3(2)^2$

A. 32

B. 36

C. 56

D. 64

3. Simplify: $\frac{(10 - 6)^2}{2(4)}$

16
8 = 2

A. 1

B. 2

C. 4

D. 8

4. Simplify the expression: $6(9 - 3) \div 3(4)$

A. 3

B. 12

C. 15.5

D. 48

5. Simplify: $3(2^3 - 2^2) + 3(4)$

A. 18

B. 24

C. 36

D. 42

6. Name the operation that is performed first.

$$4 + 3(2 + 1)^3$$

A. addition

B. exponent

C. parentheses

D. multiplication

7. Simplify: $\frac{4(15 - 3)}{2(3)} + 5$

A. 4

B. 8

C. 13

D. 20

Short Response

8. **Part A.**

Name the operation that is performed first.

$$2(3^3 - 3^2)$$

Answer__________

Part B.

Perform the order of operations on the above expression. Show your work.

Answer__________

NEGATIVE AND ZERO EXPONENTS

So far, all of the exponents we have used have been positive. But as we learned in the integer chapter, numbers can be positive, negative, and include zero. So what happens if the exponent is a zero or negative?

Look at the pattern below:

$10^2 = 100$

÷ 10

$10^1 = 10$

÷ 10

$10^0 = 1$

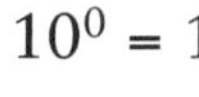

÷ 10

$10^{-1} = \frac{1}{10}$

÷ 10

$10^{-2} = \frac{1}{100}$

Patterns to Notice

The exponents are decreasing: 2, 1, 0, –1, –2

To go from one answer to the next, you divide by 10.

> RULE 1: Any nonzero number to the zero power is equal to 1. $x^0 = 1$, where x is any number except zero.
>
> RULE 2: Any nonzero number to a negative exponent equals 1 divided by the number to the positive exponent.
>
> $5^{-2} = \frac{1}{5^2} = \frac{1}{25}$ $\qquad$ $2^{-3} = \frac{1}{2^3} = \frac{1}{8}$
>
> $x^{-a} = \frac{1}{x^a}$, where x is any number except zero, and a is a positive integer.

Examples

1. Write 10^{-5} using a positive exponent.

Solution: $10^{-5} = \frac{1}{10^5}$

2. Evaluate: $4(5^0)$

Solution: $4(5^0)$

$4(1)$

4

3. Write 10^{-2} as a decimal.

Solution: $10^{-2} = \frac{1}{10^2} = \frac{1}{100} = 0.01$

4. .0001 = ________

A. 10^{-3}

B. 10^{-4}

C. 10^{-5}

D. 10^{-6}

Solution: B

The decimal .0001 is read as one ten-thousandths. As a fraction it is written $\frac{1}{10,000}$.

Since there are 4 zeros, the exponent on the 10 is 4. Because the number was a decimal, the exponent is negative.

TEST YOUR SKILLS (For answers, see page 240.)

1. Evaluate: 10^3

A. 1

B. 10

C. 100

D. 1000

2. Evaluate: 10^{-2}

A. 100

B. −100

C. $\frac{1}{100}$

D. $\frac{-1}{100}$

3. Evaluate: 5^0

A. 0

B. 1

C. 5

D. 50

4. $(-10)^4 =$ __________

A. 1000

B. −10,000

C. 10,000

D. $\frac{1}{10,000}$

5. $10^{-3} =$ __________

A. .01

B. .001

C. .0001

D. −1000

6. Evaluate: $2(3^0)$

A. 0

B. 1

C. 2

D. 3

Short Response

7. Order these numbers from least to greatest.

$$10^{-2} \quad 100 \quad .0001 \quad 10 \quad \frac{1}{10}$$

Show your work.

$10^{-2} = \frac{1}{100} = .01$ $\frac{1}{10} = .1$

$10^{-2} = .01$

.0001

Answer __________

SQUARE ROOTS AND ESTIMATING SQUARE ROOTS

You have learned how to use the symbols +, –, ×, and ÷. Now let's introduce a new symbol.

This symbol is called square root or the radical sign.

Square Root	
$\sqrt{1}$	= 1
$\sqrt{4}$	= 2
$\sqrt{9}$	= 3
$\sqrt{16}$	= 4
$\sqrt{25}$	= 5
$\sqrt{36}$	= 6
$\sqrt{49}$	= 7
$\sqrt{64}$	= 8
$\sqrt{81}$	= 9
$\sqrt{100}$	= 10
$\sqrt{121}$	= 11
$\sqrt{144}$	= 12
$\sqrt{169}$	= 13
$\sqrt{196}$	= 14
$\sqrt{225}$	= 15

A perfect square is a whole number that can be named as a product of a number with itself.

Example: $9 = 3 \times 3 = 3^2$

1, 4, 9, 16, 25, 36, 49, 64, 81, and 100 are some examples of perfect squares.

Square		Perfect Squares
1^2	=	1
2^2	=	4
3^2	=	9
4^2	=	16
5^2	=	25
6^2	=	36
7^2	=	49
8^2	=	64
9^2	=	81
10^2	=	100

Examples

1. What is the square root of 49?

   Solution: The square root of 49 means $\sqrt{49}$. The square root of 49 is 7.

2. Find the square of 3.

   Solution: The square of 3 means 3^2. The square of 3 is 9.

3. Which whole number is closest to $\sqrt{167}$?

   Solution: The $\sqrt{144} = 12$ and the $\sqrt{169} = 13$. Since $\sqrt{167}$ is closer to $\sqrt{169}$, the $\sqrt{167}$ is closest to 13.

TEST YOUR SKILLS (For answers, see page 241.)

1. Find: $\sqrt{36}$

   A. 6
   B. 18
   C. 36
   D. 72

2. Which of the following numbers is *not* a perfect square?

   A. 25
   B. 36
   C. 55
   D. 64

3. Find the square root of 121.

A. 10

B. 11

C. 12

D. 13

4. The area of a square is 144 square units. What is the length of one of its sides?

A. 12

B. 36

C. 60

D. 72

5. Between which two whole numbers is $\sqrt{80}$ on a number line?

A. 6 and 7

B. 7 and 8

C. 8 and 9

D. 9 and 10

6. The value of $\sqrt{123}$ is between what two consecutive whole numbers?

A. 10 and 11

B. 11 and 12

C. 12 and 13

D. 13 and 14

Short Response

7. A list of numbers is given below.

4 6 9 25 48 121

Part A.

Which of the numbers listed above are perfect squares?

Answer ________

Part B.

Explain your reasoning in Part A.

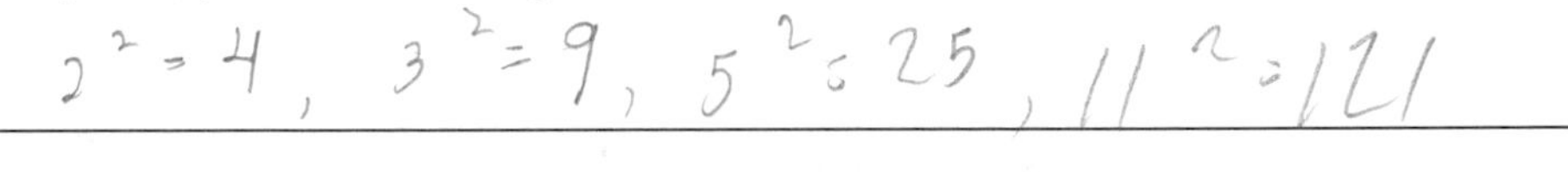

____

____

Part C.

Give an example of a number greater than 50 that is *not* a perfect square.

Answer ________

SUBSETS OF NUMBERS

- What type of sport do you play? Basketball, football, soccer, or lacrosse?
- What kind of coins are in your pocket? Quarters, nickels, dimes, or pennies?
- What grade are you in? 6th, 7th, or 8th?

We have names for many different things and we group objects and items according to certain characteristics. The same is true for grouping numbers. Names have been given to different "families" of numbers to help us group them.

There are two major families of numbers: rational and irrational numbers.

Listed below are the group names and examples of the numbers that belong to them.

Rational Numbers

- A number that can be written as a fraction
- A decimal that stops/ terminates
- A decimal that has a repeating pattern

Examples:

$\frac{3}{4}$ .6666666 ...

1.25 $\sqrt{25}$

.34 ...

Rational numbers can be broken down into more specific families.

Natural/Counting Numbers
{1, 2, 3, 4 ...}

Whole Numbers
{0, 1, 2, 3, 4, ...}

Integers
{... −3, −2, −1, 0, 1, 2, 3, ...}

Irrational Numbers

"Messy Numbers"

- A number that cannot be written as a fraction
- A decimal that goes on forever/nonterminating
- A decimal that does not repeat

Examples:

π $\sqrt{13}$

.12345 ...

$-\sqrt{17}$

TEST YOUR SKILLS (For answers, see page 242.)

1. Which of these is *not* a whole number?

   A. -5

   B. 7

   C. $\frac{10}{2}$

   D. $\sqrt{16}$

2. Which of these numbers is irrational?

   A. $\frac{1}{2}$

   B. $\sqrt{3}$

   C. 1.25

   D. -4

3. Which is the set of whole numbers?

   A. $\{\ldots -1, 0, 1 \ldots\}$

   B. $\{1, 2, 3 \ldots\}$

   C. $\{0, 1, 2, 3, \ldots\}$

   D. $\{2, 4, 6, 8\}$

4. Which of these numbers is an integer?

   A. 7.3

   B. $-\frac{1}{3}$

   C. -5

   D. $\frac{1}{2}$

5. Which number in this list is rational?

$$0.65,\ \sqrt{12},\ \pi,\ -\sqrt{18}$$

A. 0.65
B. $\sqrt{12}$
C. π
D. $-\sqrt{18}$

6. Which number is *not* a natural number?

A. 0
B. 1
C. $\sqrt{4}$
D. 3

7. What type of number is $\sqrt{5}$?

A. natural
B. integer
C. rational
D. irrational

8. What type of number is $-1\frac{2}{5}$?

A. integer
B. rational
C. whole
D. irrational

Short Response

9. Ms. Smith wrote the following numbers on the board:

$$-3 \quad \sqrt{16} \quad \pi \quad \sqrt{13} \quad \frac{1}{2} \quad -\sqrt{15}$$

Part A.

Using the numbers above, fill in the table by writing the numbers in the appropriate column.

RATIONAL	IRRATIONAL
$\sqrt{16}$	$-\sqrt{15}$
$\frac{1}{2}$	$\sqrt{13}$
-3	π

Part B.

Is the number 3.122333444455555 ... rational or irrational? Explain your reasoning.

Irrational, because it is repeating.

RATIONAL AND IRRATIONAL NUMBERS ON A NUMBER LINE

Has your teacher ever made you line up for an activity in class? Maybe you line up by your birth date or house number. By doing this, you are putting yourself in numerical order.

An easy way to compare numbers is to graph the numbers on a number line. This will help you see which numbers are the least and the greatest.

Examples

Use the number line below to answer the following questions.

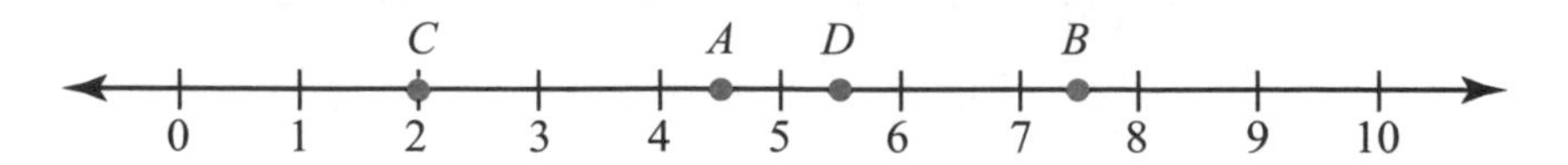

1. Which letter shows the best approximation for $\sqrt{17}$?

Solution: <u>A</u>

The $\sqrt{17}$ is between the $\sqrt{16}$ and $\sqrt{25}$. Since $\sqrt{16} = 4$ and $\sqrt{25} = 5$, the $\sqrt{17}$ is between 4 and 5.

2. Which letter shows the best approximation for $\sqrt{60}$?

Solution: <u>B</u>

The $\sqrt{60}$ is between the $\sqrt{49}$ and $\sqrt{64}$. Since $\sqrt{49} = 7$ and $\sqrt{64} = 8$, the $\sqrt{60}$ is between 7 and 8.

Example

Use a number line to order the numbers $\sqrt{23}$, 1.7, π, and –3.1 from least to greatest.

Solution: Numbers need to be in the same form to compare them. In this case, change every number to a decimal.

$\sqrt{23}$	The $\sqrt{16} = 4$ and $\sqrt{25} = 5$, so the $\sqrt{23}$ is between 4 and 5.
1.7	1.7 is a decimal between 1 and 2.
π	π is approximately 3.14, so it is between 3 and 4.
–3.1	–3.1 is a decimal between –3 and –4.

The numbers from least to greatest are –3.1, 1.7, π, $\sqrt{23}$.

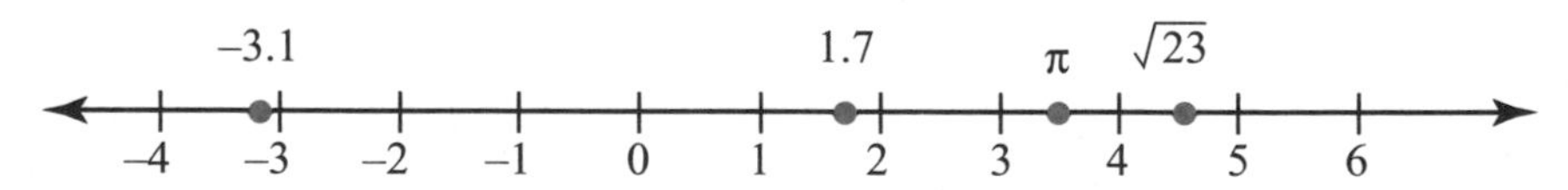

TEST YOUR SKILLS (For answers, see page 243.)

1. Which letter shows the $\sqrt{49}$ on the number line?

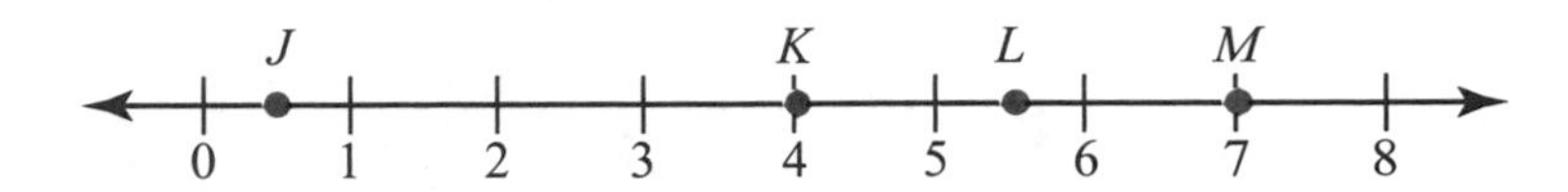

A. J
B. K
C. L
D. M

2. Which point is *not* graphed on the number line?

A. π
B. $\frac{1}{2}$
C. $\sqrt{9}$
D. $\sqrt{25}$

3. Which number line shows $-\pi$ graphed correctly?

A. 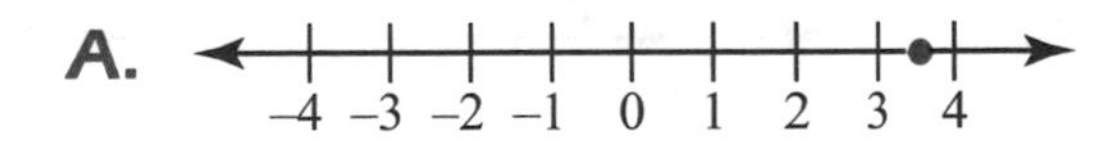

B. –4 –3 –2 –1 0 1 2 3 4

C. –4 –3 –2 –1 0 1 2 3 4

D. –4 –3 –2 –1 0 1 2 3 4

4. Which number line shows $\sqrt{50}$ graphed correctly?

A. 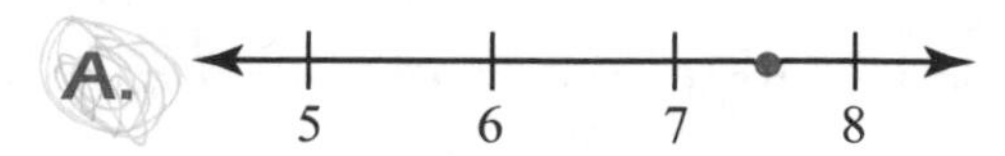

B. 50 60 70 80

C. 25 26 27 28

D. 5 6 7 8

5. The point graphed on the following number line could represent which number?

A. -3.2

B. $\sqrt{4}$

C. $\sqrt{12}$

D. $\sqrt{25}$

6. Use the number line below to order $\sqrt{5}$, -0.5, and $\frac{8}{2}$ from least to greatest.

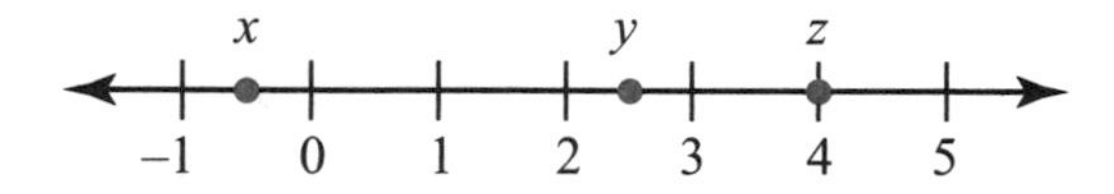

A. $-0.5, \frac{8}{2}, \sqrt{5}$

B. $\frac{8}{2}, \sqrt{5}, -0.5$

C. $\sqrt{5}, \frac{8}{2}, -0.5$

D. $-0.5, \sqrt{5}, \frac{8}{2}$

7. Use the number line below to order $-\frac{1}{4}$, $\sqrt{16}$, $\sqrt{3}$, and 5.75 from greatest to least.

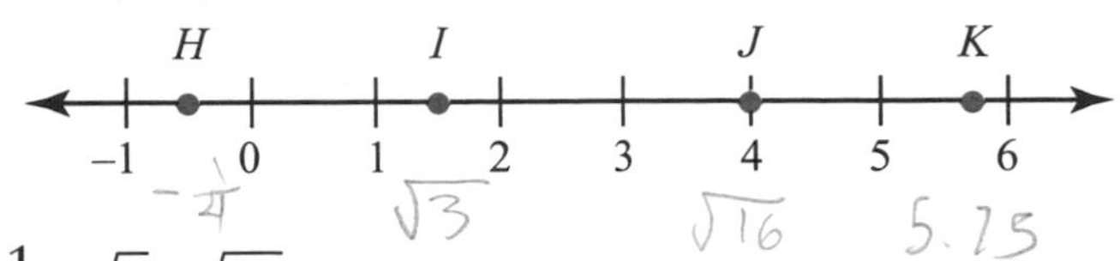

A. $-\frac{1}{4}$, $\sqrt{3}$, $\sqrt{16}$, 5.75

B. 5.75, $\sqrt{16}$, $\sqrt{3}$, $-\frac{1}{4}$

C. $\sqrt{3}$, $-\frac{1}{4}$, 5.75, $\sqrt{16}$

D. $\sqrt{16}$, 5.75, $-\frac{1}{4}$, $\sqrt{3}$

Short Response

8. A list of numbers is given below.

$$-\frac{1}{2},\ \sqrt{9},\ -\sqrt{4},\ \pi,\ \frac{1}{4}$$

Part A.

Graph the numbers on the number line below. Show your work.

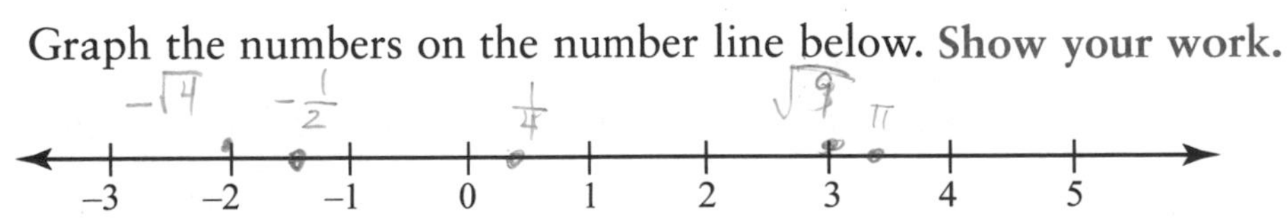

Part B.

List the numbers from greatest to least.

Answer ________

REPEATING AND NONREPEATING DECIMALS

Let's review our number system:

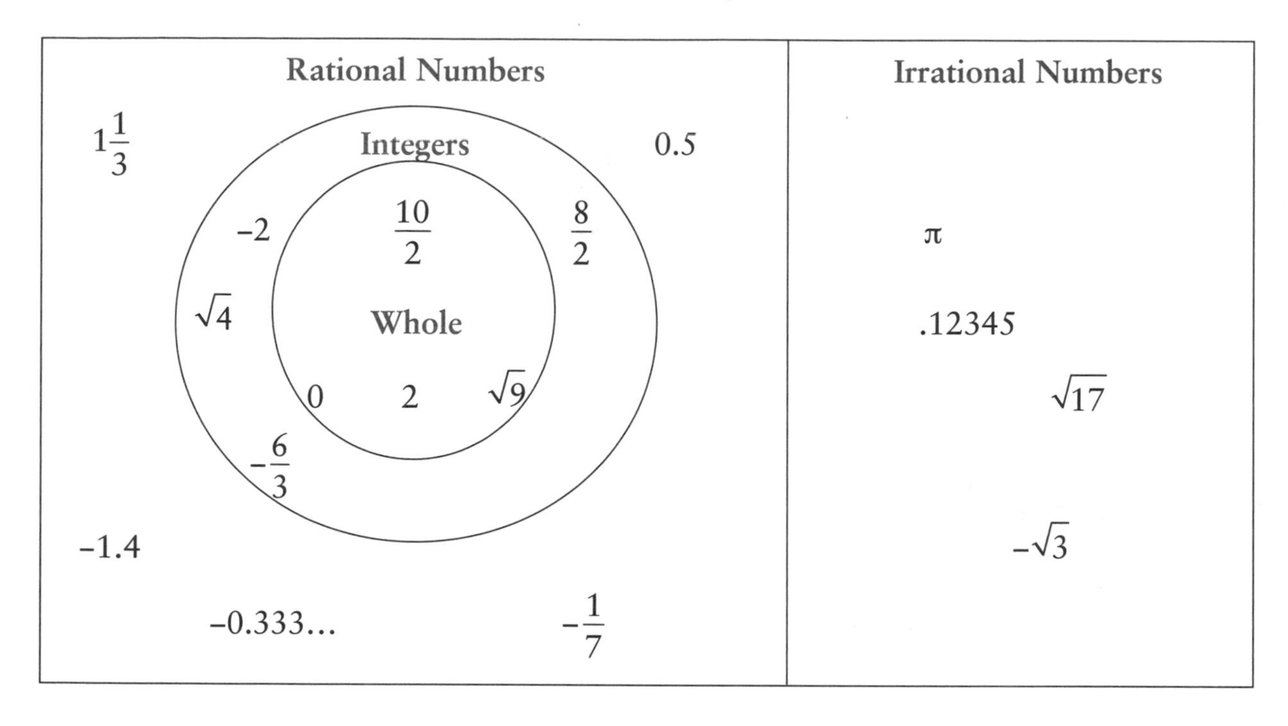

Just like there are two types of numbers, rational and irrational, there are also two types of decimals.

DECIMALS

TERMINATING
Decimals that end, stop, "terminate"

Examples:
2.5
54.623
$\sqrt{16} = 4$
$\frac{1}{2} = 0.5$

NONTERMINATING
Decimals that go on forever

Repeating
Decimals that have a repeating pattern

*Examples:
4.12121212...
$\frac{1}{3} = 0.33333\ldots$

*Note:
These are rational numbers

Nonrepeating
Decimals that have no pattern

*Examples:
$\pi = 3.14159265\ldots$
$\sqrt{12}$

*Note:
These are irrational numbers

Examples

1. Which description best describes a repeating decimal?

   A. a terminating decimal

   B. a decimal that has no pattern

   C. a decimal that has a pattern and can be written as a fraction

   D. an irrational number

   Solution:

   C A repeating decimal is a nonterminating decimal that has a repeating pattern. It can also be written as a fraction.

   Example: $\frac{1}{3} = 0.33333\ \ldots$

2. Which of the following is a nonterminating decimal?

   A. $\frac{1}{2}$

   B. $\sqrt{25}$

   C. -5

   D. $\sqrt{3}$

   Solution:

   D Nonterminating means the decimal does not end. $\frac{1}{2} = 0.5$, $\sqrt{25} = 5$, and -5 are all numbers that end. The $\sqrt{3}$ is an irrational number that goes on forever.

TEST YOUR SKILLS (For answers, see page 244.)

1. Which of these numbers is a repeating decimal?

   A. $\frac{1}{2}$

   B. $\frac{1}{4}$

   C. $\frac{1}{10}$

   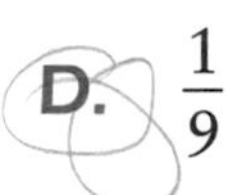

   D. $\frac{1}{9}$

2. What type of number is 0.75?

   A. terminating and rational

   B. whole

   C. integer

   D. terminating and irrational

3. When written in decimal form, which of these is a terminating decimal?

   A. $\frac{1}{3}$

   B. π

   C. $\frac{4}{5}$

   D. $\sqrt{13}$

4. When written in decimal form, which of these is a nonterminating decimal?

A. $\sqrt{3}$

B. $1\frac{1}{2}$

C. $\frac{1}{4}$

D. $\sqrt{36}$

5. Which type of number is a nonrepeating, nonterminating decimal?

A. -3π

B. $\frac{1}{9}$

C. $\frac{1}{8}$

D. $\frac{1}{5}$

6. What type of number is $\frac{9}{11}$?

A. whole

B. natural

C. rational and repeating

D. irrational and repeating

7. What type of number is $\sqrt{49}$?

A. rational and terminating

B. irrational and terminating

C. rational and nonterminating

D. irrational and nonterminating

Short Response

8. Reagan and Miles are partners in math class. They are having trouble with the following problems, so they ask you for help.

Part A.

Is $\frac{2}{5}$ a terminating or nonterminating decimal? Show your work and explain your reasoning.

____

____

____

Part B.

Is $\frac{1}{6}$ a terminating or nonterminating decimal? Show your work and explain your reasoning.

____

____

____

JUSTIFY REASONABLENESS USING ESTIMATION

You are at the mall shopping. You want to buy a shirt for $19.99, a CD for $9.99, and a pair of shoes for $29.99. Estimate how much money you will spend on all three items.

To estimate, round each amount to the nearest dollar. Then add the prices of all three items.

$19.99 → $20.00
$9.99 → $10.00
$29.99 → $30.00

To buy all three items, you will spend about $60.00

You can estimate when you are adding, subtracting, multiplying, or dividing.

Examples

1. You and your friends are going to a concert. You spent $159.96 on four concert tickets. About how much money does each friend owe you for a ticket?

Solution: Round the cost of the tickets to the nearest dollar.
$159.96 rounds to $160.
Divide by the number of tickets. $160 ÷ 4 = $40.
Each friend owes you about $40.

2. In a local middle school, there are 213 sixth graders and 302 seventh graders. If there are 746 students in the sixth, seventh, and eighth grades, about how many students are there in the eighth grade?

Solution: To find the number of eighth graders, add up the sixth and seventh graders and subtract this number from the total number of students in the middle school. Use rounding to estimate the number of students.

Step 1: Round the number of sixth and seventh graders.
213 rounds to 200 sixth graders.
302 rounds to 300 seventh graders.

Step 2: Add the sixth and seventh graders together.
200 + 300 = 500

Step 3: Round the total number of students in the middle school.
746 students rounds to about 750 students.

Step 4: To find the number of eighth graders, subtract 750 – 500.
There are about 250 eighth graders in the middle school.

TEST YOUR SKILLS (For answers, see page 244.)

1. Your family drove 23.4 miles on Friday, 28.9 miles on Saturday, and 33.1 miles on Sunday. About how many miles did your family travel in 3 days?

   A. 70 miles

   B. 85 miles

   C. 100 miles

   D. 120 miles

2. Ericka bought four t-shirts at $7.99 each. About how much did Ericka spend on t-shirts?

   A. $2

   B. $12

   C. $24

   D. $32

3. Your parents' car has a 15-gallon gas tank. If gas costs $3.23 a gallon, about how much money will it cost your parents to fill up their gas tank?

   A. $18

   B. $25

   C. $32

   D. $48

4. For lunch, you bought taco salad for $2.25 and an ice cream for $0.99. Your best friend Kyle bought the chicken sandwich for $2.99 and a cookie for $1.99. About how much more money did Kyle spend on lunch than you?

   A. $1

   B. $2

   C. $3

   D. $4

5. What is a reasonable estimate for 112.34 ÷ 3.89?

   A. 28

   B. 40

   C. 50

   D. 82

6. Your math teacher bought a big bucket of candy to hand out in class. There are 296 pieces of candy in it. Your class has 27 students. If your teacher passes out all of the candy in the bucket, about how many pieces of candy would each student in the class get, if no one was absent?

   A. 5

   B. 10

   C. 15

   D. 20

7. On a family trip, you drove 7,466 miles in 14 days. About how many miles did your family drive each day?

   A. 200 miles

   B. 350 miles

   C. 500 miles

   D. 700 miles

8. What is a reasonable estimate for 21.16 × 5.79?

   A. 75

   B. 90

   C. 126

   D. 140

Short Response

9. Brenda wants to buy 19 packages of plates for a graduation party. The plates cost $1.89 per package.

Part A.

If Brenda has $30, does she have enough money to buy all 19 packages of plates? Using estimation, show your work.

Answer __________

Part B.

If Brenda only has $30, about how many packages of plates would she be able to buy if they cost $1.89 per package? Using estimation, show your work.

Answer __________

GREATEST COMMON FACTOR AND LEAST COMMON MULTIPLE

A factor is a number that divides evenly into another number.

What are the factors of 12? The factors of 12 are the numbers that will divide evenly into 12.

An easy way to list the factors of a number is to make a table of the numbers that multiply to 12.

12		
1	12	How do you know when you have all the numbers?
2	6	You look for a set of repeated numbers.
3	4	
4	3 ←	(repeated number) Since 4 × 3 shows up a second time, you are finished.

So the factors of 12 are 1, 2, 3, 4, 6, and 12.

Example

Find the factors of 18.

18	
1	18
2	9
3	6
4	–
5	–
6	3 (repeated number)

So the factors of 18 are 1, 2, 3, 6, 9, and 18.

GREATEST COMMON FACTOR (GCF)

The GCF of two or more numbers is the largest number that is a factor of each number.

Example

Find of the GCF of 45 and 60.

45	
1	45
2	–
3	(15)
4	–
5	9
6	–
7	–
8	–
9	5 (repeat)

60	
1	60
2	30
3	20
4	(15)
5	12
6	10
7	–
8	–
9	–
10	6 (repeat)

To find the GCF, look for the biggest number that appears in BOTH tables. It does not matter if the number shows up on the left or right side of the table, it just needs to appear in both tables.

The highest number in both tables is 15. *So the GCF of 45 and 60 is 15.*

MULTIPLES

Multiples are the numbers that result from multiplying a number by the counting numbers.

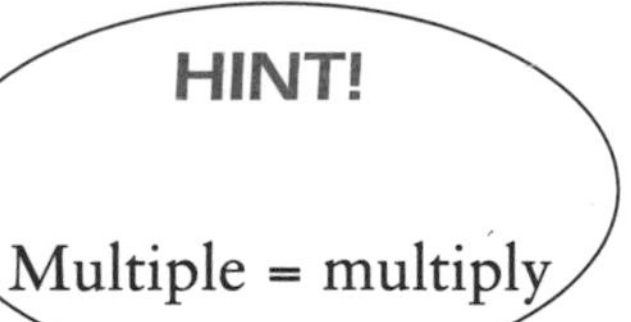

Example

What are the multiples of 4?

The multiples of 4 are the numbers you get when you multiply 4 by 1, 2, 3, ...

$4 \times 1 = 4$
$4 \times 2 = 8$ So the multiples of 4 are 4, 8, 12, 16, ...
$4 \times 3 = 12$
$4 \times 4 = 16$

Example

What are the first four multiples of 6?

$6 \times 1 = 6$
$6 \times 2 = 12$ So the first four multiples of 6 are 6, 12, 18, and 24.
$6 \times 3 = 18$
$6 \times 4 = 24$

LEAST COMMON MULTIPLE (LCM)

The LCM of two numbers is the smallest number that is a multiple of both numbers.

Example

Find the LCM of 12 and 16.

12	12, 24, 36, (48)
16	16, 32, (48)

List the multiples for each number until you find the smallest number that appears in both lists.

The LCM of 12 and 16 is 48.

TEST YOUR SKILLS (For answers, see page 245.)

1. 16 is the greatest common factor of what two numbers?

    A. 12 and 24

    B. 32 and 48

    C. 36 and 42

    D. 48 and 60

2. Find the greatest common factor of 35, 49, and 84.

    A. 5

    B. 6

    C. 7

    D. 9

3. What is the LCM of 8 and 20?

    A. 4

    B. 24

    C. 40

    D. 160

4. What is the least common multiple of 4, 8, and 12?

    A. 4

    B. 12

    C. 24

    D. 48

5. 56 is the least common multiple of which three numbers?

    A. 2, 6, 12

    B. 2, 7, 8

    C. 2, 8, 10

    D. 3, 5, 6

6. Your dad needs to tile the wall behind the sink in your bathroom. The size of the wall is 40 inches by 16 inches. If he has 8-inch square tiles, how many tiles does he need to cover the wall?

A. 7

B. 10

C. 12

D. 15

7. On the school lunch menu, every sixth day they serve hot dogs and every eighth day they serve tacos. On which day will they serve hot dogs and tacos?

A. 2

B. 12

C. 24

D. 48

Short Response

8. You are helping your mom put gift bags together for an upcoming party. She wants to put lollipops in each gift bag. She bought a box of 36 orange lollipops and another box of 42 cherry lollipops.

Part A.

What is the greatest number of lollipops you can put in each bag so that each person gets the same number of cherry and orange lollipops? Show your work.

Answer __________

Part B.

If your mom wants to put 3 lollipops of any color in each bag, how many gift bags would you have? **Show your work.**

Answer __________

PRIME FACTORIZATION

A **prime number** is a whole number greater than 1 that only has factors of 1 and itself.

2 is a prime number because its factors are 1 and 2.
3 is a prime number because its factors are 1 and 3.

A few of the prime numbers include:
2, 3, 5, 7, 9, 11, 13, 17, 19, ...

A **composite number** is a whole number greater than 1 that has more than two factors.

8 is a composite number because its factors are 1, 2, 4, and 8.

To find the prime factorization of a number, you make a factor tree. A **factor tree** is a diagram that can be used to write the prime factorization of a number.

Example

Find the prime factorization of 108.

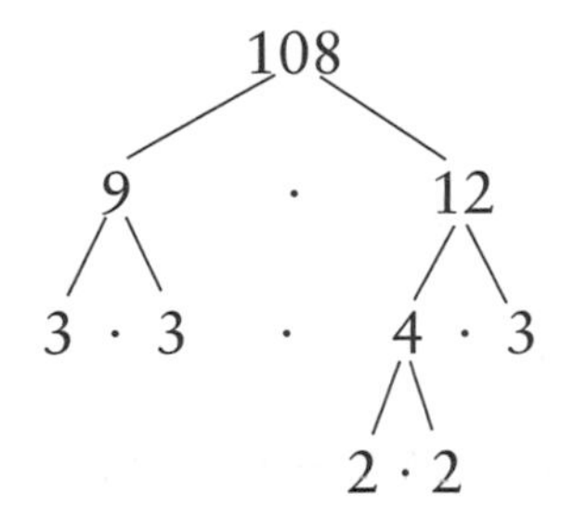

The prime factorization of 108 is $2 \times 2 \times 3 \times 3 \times 3$.
The prime factorization of 108 written in exponential form is $2^2 \times 3^3$.

TEST YOUR SKILLS (For answers, see page 246.)

1. The prime factorization of 36 is _________.

   A. $1 \times 2^2 \times 3^2$

   B. $2^2 \times 3^2$

   C. $3^2 \times 4$

   D. 9×4

2. Which of these numbers is prime?

   A. 1

   B. 4

   C. 9

   D. 13

3. Which of these numbers is composite?

   A. 2

   B. 11

   C. 22

   D. 29

4. Find the prime factorization of 120.

   A. $2 \times 3 \times 4 \times 5$

   B. $2^2 \times 3 \times 5$

   C. $2^3 \times 3 \times 5$

   D. 10×12

5. What number has a prime factorization of $2^3 \cdot 3^2 \cdot 5$?

   A. 180

   B. 240

   C. 360

   D. 408

6. Which number has a prime factorization of $3^2 \cdot 13$?

A. 36

B. 78

C. 84

D. 117

Short Response

7. Find the prime factorization of 720.

Part A.

Show your work.

Answer __________

Part B.

Write your answer to Part A in exponential form.

Answer __________

8. On his math test, Chris was asked to find the prime factorization of 60. The following is his factor tree:

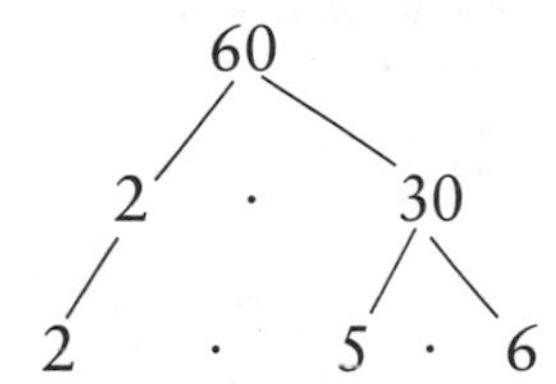

Did Chris do this problem correctly? Using what you know about prime factorization, explain your reasoning.

SCIENTIFIC NOTATION AND TRANSLATING SCIENTIFIC NOTATION INTO STANDARD FORM

Do you have a friend who's name might be Samantha, but she likes to be called Sam? Or do you know someone named Jonathan, who wants to be called John? People tend to shorten their names or give themselves nicknames. Likewise, when dealing with very large numbers it is often easier to write the number in a shorthand form. This simplified form is called scientific notation.

SCIENTIFIC NOTATION

A number like 230,000,000,000,000,000,000,000,000 is written as 2.3×10^{26}. So how do you write a number in scientific notation?

It looks like

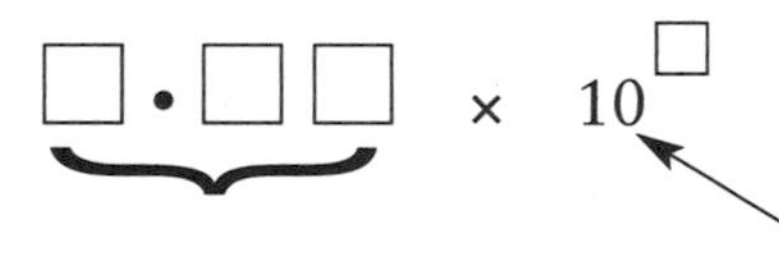

The first factor is greater than or equal to 1 and less than 10.

Examples: 5, 6.05, 7.023

The second factor is the number 10 raised to an exponent. The exponent equals the number of places to move the decimal point.

Examples of numbers written in scientific notation	Examples of numbers NOT written in scientific notation
4×10^8	45.6×10^3
6.03×10^4	0.4×10^{11}
5.34×10^2	300×10^2
Notice that all of the first factors are greater than 1 and less than 10!	Notice that all of the first factors are greater than 10 or less than 1!

Examples

1. Write the number 412,000 in scientific notation.

   Solution: 4.12×10^5

2. Write the number 8000 in scientific notation.

   Solution: 8×10^3

WRITING NUMBERS IN STANDARD FORM

If a number is written in scientific notation, it is sometimes helpful to write the number out (standard form).

Examples

1. Write the number 6.2×10^5 in standard form.

   Solution: The exponent (5) tells you how many times to move the decimal place.

$$6.2 \times 10^5 \rightarrow 6 \cdot 2\ 0\ 0\ 0\ 0 = 620{,}000$$

2. Write the number 4.02×10^3 in standard form.

$$4.02 \times 10^3 \rightarrow 4 \cdot 0\ 2\ 0 = 4020$$

SCIENTIFIC NOTATION WITH NEGATIVE EXPONENTS

Very small numbers (decimals) can also be written in scientific notation.

Examples

1. Write the number 0.0045 in scientific notation.

Solution: 4.5×10^{-3}

Notice that the first factor is still greater than 1 and less than 10.
The decimal is negative because you are moving the decimal three places to the *left*.

2. Write the number 0.000000026 in scientific notation.

Solution: 2.6×10^{-8}

3. Write the number 3.02×10^{-5} in standard form.

Solution: 0.0000302

> REMEMBER: If the exponent is positive, move the decimal point to the *right*.
> If the exponent is negative, move the decimal point to the *left*.

TEST YOUR SKILLS (For answers, see page 247.)

1. Write the number 214,000 in scientific notation.

A. 2.14×10^3

B. 21.4×10^4

C. 2.14×10^5

D. 214×10^3

2. Write the number 300 in scientific notation.

A. 3×10^2

B. 3×10^3

C. 3×100

D. 30×10

3. Write the number .004 in scientific notation.

A. 0.4×10^{-2}

B. $4 \times .001$

C. 4×10^{-3}

D. 4×10^3

4. Express the number 3.45×10^5 in standard form.

A. .00345

B. 3.4500000

C. 34,500

D. 345,000

5. The Empire State Building is 1.25×10^3 feet tall. Write this number in standard form.

A. 1.25 feet

B. 125 feet

C. 1,250 feet

D. 12,500 feet

6. The number of students in a middle school are given in the table below:

Grade Level	Number of Students
6th	542
7th	603
8th	588

Which of the following shows the number of students in seventh grade written in scientific notation?

A. 5.42×10^2

B. 6×10^2

C. 6.03×10^2

D. 603

7. Which number is *not* written in scientific notation?

A. 1.4×10^2

B. 4.035×10^3

C. 6×10^6

D. 23.6×10^2

Short Response

8. Tina and her friend Emily are checking over their scientific notation homework together. They can't seem to decide who has the correct answer for the following problem:

Write the number 54,500,000 in scientific notation.

Tina thinks the answer is 545×10^5 and Emily thinks the answer is 5.45×10^7.

Part A.

Who is correct?

Answer __________

Part B.

Explain your reasoning.

____

____

____

COMPARING NUMBERS IN SCIENTIFIC NOTATION

In order to compare numbers written in scientific notation, you could write each number out in standard form. Or, you could use what you know about scientific notation and compare the factors and the exponents.

Which Is Greater?

CASE 1

3.4×10^5 $\quad$ 6.5×10^5

Since the exponents are the same, compare the first factors of each number.

Since 6.5 is greater than 3.4, 6.5×10^5 is the greater number.

CASE 2

2.98×10^8 $\quad$ 4.7×10^4

Since the powers of 10 are different, compare the exponents of each number.

Since 8 is greater than 4, 2.98×10^8 is the greater number.

You can use these rules to determine which number written in scientific notation is the largest or the smallest.

To compare numbers in scientific notation:

- If the powers of 10 are the *same*, compare the first factors of each number.
- If the powers of 10 are *different*, compare the exponents of each number.

TEST YOUR SKILLS (For answers, see page 248.)

1. Which number is the greatest?

   A. 3.6×10^4

   B. 4.5×10^4

   C. 36,500

   D. 42,000

2. Which number is the least?

   A. 1.7×10^7

   B. 2.8×10^8

   C. 3.2×10^7

   D. 6.2×10^8

3. Which list of numbers is in order from least to greatest?

   A. 2.8×10^3, 3.4×10^3, 5.3×10^2, 6.1×10^4

   B. 3.4×10^3, 5.3×10^2, 6.1×10^4, 2.8×10^3

   C. 5.3×10^2, 2.8×10^3, 3.4×10^3, 6.1×10^4

   D. 6.1×10^4, 3.4×10^3, 5.3×10^2, 2.8×10^3

4. The populations of four countries as of 2007 are given in the table below:

Country	Population
United States	3.01×10^8
Canada	3.34×10^7
Japan	1.275×10^8
Germany	8.24×10^7

Source: U.S. Census Bureau, International Database

Which of the following lists the populations in order from least to greatest?

A. Canada, Japan, Germany, United States

B. United States, Japan, Canada, Germany

C. Japan, Germany, United States, Canada

D. Canada, Germany, Japan, United States

5. The mass of four of the planets are given below.

Planet	Mass in kg
Earth	5.98×10^{24}
Mercury	3.3×10^{25}
Venus	4.87×10^{24}
Mars	6.5×10^{25}

Source: The World Almanac

Which planet has the smallest mass?

A. Earth

B. Mercury

C. Venus

D. Mars

Short Response

6. In social studies, you are studying different facts about the Great Lakes. Your teacher knows that in math class you have learned about scientific notation. Your social studies teacher has given you the following index cards to use for your homework assignment. Each card lists the area of a Great Lake in square kilometers.

Lake Erie	Lake Huron	Lake Superior	Lake Michigan	Lake Ontario
2.57×10^4	59,600	8.2×10^4	57,800	1.896×10^4

Source: The World Almanac

Part A.

Using the information written on the cards, list the areas of the Great Lakes from smallest to largest.

Show your work.

Answer __________

Part B.

Which Great Lake has the largest area?

Answer __________

Chapter 2

ALGEBRA

EVALUATING ALGEBRAIC EXPRESSIONS

Algebraic Expression

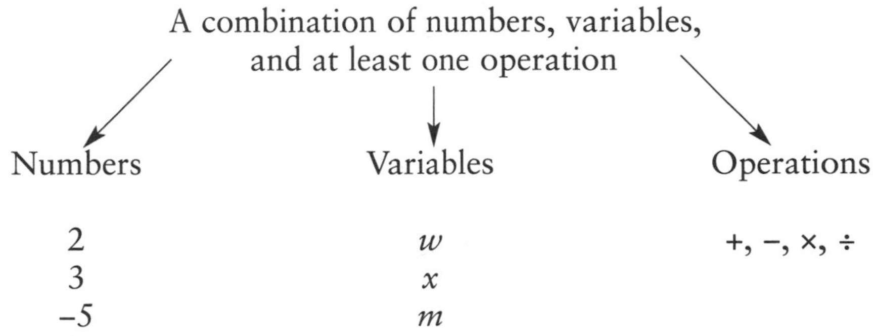

Examples: $2w$
$4x + 3$
$m + 2n - 5$

It is important to know the words that are used when using algebraic expressions.

Here is a quick review.

Vocabulary Review

Variable	A letter used to represent a number in an expression or an equation	$\underline{x}$, $2\underline{a}$, $3\underline{x} + \underline{y}$
Coefficient	The number in front of the variable	$\underline{3}a$
Constant	A value that does not change; a number	$2x + \underline{5}$

We use algebraic expressions in math to solve for something that is unknown.

Examples

1. Alyssa is three more than five times her brother Derek's age. The expression $5d + 3$ can be used to find Alyssa's age. Let d represent Derek's age. If Derek is 5 years old, how old is Alyssa?

   Solution:

   $5d + 3$ is an algebraic expression. We know that d represents Derek's age. Substitute 5 in for d to solve for Alyssa's age.

   $5d + 3$
   $5(5) + 3$
   $25 + 3$
   28

   Alyssa is 28 years old.

2. What is the value of the expression $3x^2 - 2y + 5$ if $x = 4$ and $y = 2$?

   To **evaluate** an expression means to find the value of the expression. To help you remember this, the word "valu" is in evaluate.

   Solution:

Rewrite the original problem.	$3x^2 - 2y + 5$
Substitute the numbers in for x and y.	$3(4)^2 - 2(2) + 5$
First simplify the exponent.	$3(16) - 2(2) + 5$
Multiply from left to right.	$48 - 4 + 5$
Subtract.	$44 + 5$
Add.	49

   The final answer is 49.

TEST YOUR SKILLS (For answers, see page 249.)

1. Evaluate $2a + 6b - 5$ if $a = 12$ and $b = 2$.

   A. 7
   B. 17
   C. 31
   D. 40

2. Evaluate $m^3 + n^2 - 2$ if $m = 3$ and $n = 2$.

A. –5
B. 11
C. 29
D. 31

3. Evaluate $(xy)^2$ if $x = 2$ and $y = 4$.

A. 12
B. 16
C. 36
D. 64

4. Find the value of the expression $4x^2 - y$ if $x = 3$ and $y = 5$.

A. 19
B. 31
C. 44
D. 139

5. Find the value of the expression $\frac{1}{2}(5m)(n^2)$ if $m = 4$ and $n = 3$.

A. 27
B. 60
C. 90
D. 162

6. Evaluate the expression $3x + 4y$ if $x = 2$ and $y = 3.5$.

A. 15
B. 20
C. 35
D. 146

7. Which values make the expression $5xy + 3$ equal to 18?

A. $x = 2, y = 1$

B. $x = 2, y = 4$

C. $x = 3, y = 1$

D. $x = 3, y = 3$

8. The expression $\$3t + \$2p$ can be used to find the total cost of going to the movies where t represents the number of tickets purchased and p represents the number of bags of popcorn purchased. How much would it cost to go the movies if you bought 4 tickets and 2 bags of popcorn?

A. 16

B. 20

C. 22

D. 28

9. What is the value of the expression $4g^2 + \frac{h}{3}$ if $g = 2$ and $h = 12$?

A. 8

B. 12

C. 17

D. 20

Short Response

10. Kim is 4 years younger than three times Tommy's age. If t represents Tommy's age, the expression $3t - 4$ can be used to represent Kim's age.

Part A.

If Tommy is 12 years old, how old is Kim? Show your work.

Answer ____________

Part B.

If Kim were 17 years old, how old would Tommy be? Show your work.

$3t - 4 = 17$
$+4 \quad +4$
$\frac{3t}{3} = \frac{21}{3}$

Answer 7

TRANSLATING 2-STEP VERBAL EXPRESSIONS INTO EQUATIONS

Math is a language, just like English or Spanish. Languages are confusing when you don't know the vocabulary. Translating math expressions can be just as confusing. The following table will help make translating math words a lot easier!

+	−
Add Plus Sum More than Increased by	Subtract Minus Difference Less than Decreased by
×	÷
Multiply Times Product Of Factor Twice Triple	Divide Quotient

Expressions vs. Equations

An algebraic expression is a combination of variables (letters), numbers, and at least one operation.

Examples: $n + 5$
xy
$2w + 3x + 4$

Example

Write an algebraic expression for the following:

Verbal Expression		Algebraic Expression
Fifteen less than a number.	→	$N - 15$
Ten more than the product of a number and three.	→	$3x + 10$
Twelve times a number minus eight.	→	$12x - 8$

An **algebraic equation** is a mathematical sentence with an equal sign (=) in it, for example: $2x + 5 = 13$.

Example

Write an algebraic equation for the following:

Verbal Expression		Algebraic Equation
Fifteen less than a number is equal to twelve.	→	$N - 15 = 12$
Ten more than the product of a number and three is equal to twenty-two.	→	$3x + 10 = 22$
Twelve times a number minus eight is equal to four.	→	$12x - 8 = 4$

TEST YOUR SKILLS (For answers, see page 251.)

1. Which equation shows two less than six times a number is equal to ten?

$2 - 6x = 10$

A. $2 - 6x = 10$
B. $2 + 6x = 10$
C. $6x - 2 = 10$
D. $6(x - 2) = 10$

2. Which equation shows the sum of five and four times a number is equal to fourteen?

A. $9x = 14$

B. $5 + 4x = 14$

C. $5x + 4x = 14$

D. $4 + 5x = 14$

3. Which equation represents the sentence "Three more than two times a number is twelve"?

A. $3n + 2 = 12$

B. $3(n + 2) = 12$

C. $3 + 2n = 12$

D. $2(n + 3) = 12$

4. A taxi cab company charges four dollars plus thirty cents per mile. If your cab ride cost thirty-six dollars, which equation could be used to see how many miles, m, you traveled?

A. $4 + 30 + m = 36$

B. $4 + .30m = 36$

C. $.30 + 4m = 36$

D. $.30 + 4 + m = 36$

5. The product of a number and twelve is equal to the sum of thirty and twice the same number. Which equation can be used to solve for the number, x?

A. $12x = 2x + 30$

B. $12 + x = 30 + 2 + x$

C. $12 x 30 = 2x$

D. $\frac{12}{x} = 2x + 30$

6. In the fall, Drew rakes leaves for people in his neighborhood. Drew charges five dollars to rake each yard plus a two dollar clean-up fee. This fall Drew made $102. Which equation can be used to determine how many yards, y, Drew has raked this fall?

A. $2y + 5 = 102$

B. $5y + 2 = 102$

C. $2y \cdot 5 = 102$

D. $5y \cdot 2 = 102$

7. Which equation represents "twice a number plus three is equal to the sum of the same number and seven"?

A. $3x + 2 = x + 7$

B. $2x + 7 = x + 3$

C. $3x + 7 = x + 2$

D. $2x + 3 = x + 7$

8. Which equation represents "The quotient of y and eight plus fifteen equals thirty-one"?

A. $8y + 15 = 31$

B. $y + 8 + 15 = 31$

C. $8y - 15 = 31$

D. $\frac{y}{8} + 15 = 31$

Short Response

9. Phone America charges $39 per month for a cell phone, plus $0.14 per minute for each additional minute over the cell phone plan.

Part A.

Let m represent each minute over the cell phone plan. Write an expression to represent the cost of the cell phone for one month.

Answer __________

Part B.

If you used an additional 10 minutes over your cell phone plan, how much would the cell phone cost you for the month? Show your work.

Answer __________

SOLVING 2-STEP EQUATIONS

Have you ever been to the movies? It costs $6 per movie ticket and $3 for a bag of popcorn. If you have $27, how many friends could you take to the movies and buy one bag of popcorn?

You can use a **2-step equation** to help you solve this problem!

First, use what you know about translating expressions to write an equation.

Let x = number of friends

$6 per movie ticket	→	$6x$
$3 for a bag of popcorn	→	$+ 3$
You have $27	→	$= 27$
Equation	→	$6x + 3 = 27$

To solve a 2-step equation:

1. Undo add or subtract.
2. Undo multiply or divide.
3. Simplify.

Examples

1. Solve: $6x + 3 = 27$

$6x + 3 = 27$ $\quad -3 \quad -3$	Rewrite the original equation. Undo the add 3 by subtracting 3 from both sides of the equation.
$\frac{6x}{6} = \frac{24}{6}$	Undo 6 times x by dividing by 6 on both sides of the equation.
$x = 4$	Simplify.

2. Solve: $\frac{x}{2} - 5 = 3$

$\frac{x}{2} - 5 = 3$	Rewrite the original equation.
$+5 \quad +5$	Undo subtract 5 by adding 5 to both sides of the equation.
$2 \cdot \frac{x}{2} = 8 \cdot 2$	Undo divide by 2 by multiplying by 2 on both sides of the equation.
$x = 16$	Simplify.

TEST YOUR SKILLS (For answers, see page 253.)

1. Solve: $2x + 168 = 432$

A. 32

B. 48

C. 132

D. 300

2. Solve: $\frac{y}{5} - 9 = 17$

A. 5

B. 26

C. 85

D. 130

3. Solve: $\frac{m}{3} + 12 = 39$

A. 9

B. 13

C. 81

D. 153

4. Your cell phone bill was $114 this month. It costs you $0.25 per text message, t, plus $39 a month for the cell phone. Solve the equation $.25t + 39 = 114$ to find the number of text messages you sent this month.

A. 30

B. 82

C. 300

D. 417

5. Your family spent the weekend in Vermont at a ski resort. You skied the same number of hours on Friday night and Saturday. You also skied on Sunday for 3 hours. If you skied a total of 13 hours over the weekend, solve the equation $2h + 3 = 13$ to see how many hours, h, you skied on Friday and Saturday.

A. 5

B. 8

C. 10

D. 12

6. Four times a number n increased by 5 is 25. Solve the equation $4n + 5 = 25$ to find the number n.

A. 4

B. 5

C. 6

D. 7

7. Benny's Balloons charges $2 for each balloon in an arrangement. There is also a $6 fee for making the arrangement. You have $16 to spend. Use the equation $2b + 6 = 16$, where b represents the number of balloons, to find the number of balloons you can buy.

A. 5

B. 6

C. 11

D. 12

8. Eight less than the quotient of a number and 2 is equal to 10. Use the equation $\frac{x}{2} - 8 = 10$ to solve for the number x.

A. 4

B. 9

C. 13

D. 36

Short Response

9. Jenna bought four shirts at the mall for a total of $68. The tax on her shirts was $12.

Part A.

Let s represent the cost of each shirt Jenna bought. Write an equation you could use to solve for the cost of each shirt.

Answer ___________

Part B.

Solve your equation in Part A to find the cost of each shirt. Show your work.

Answer ___________

10. Anthony has asked you to help him with his solving equations homework. His teacher told him that he did not solve the following problem correctly and he needs to make corrections. Anthony is not sure what he did wrong. His work is shown below.

$$\frac{x}{2} - 5 = 9$$

$$2 \cdot \frac{x}{2} - 5 = 9 \cdot 2$$

$$x - 5 = 18$$

$$\underline{\quad +5 \qquad +5}$$

$$x = 23$$

Part A.

What did Anthony do wrong when he was solving the equation?

___

___

___

Part B.

Show Anthony how to correctly solve the problem. Show your work.

Answer ______

PROPORTIONS

A proportion is an equation that shows two ratios as equivalent. We can use proportions to help us solve for an unknown variable.

Examples

1. Determine if the ratios $\frac{2}{4}$ and $\frac{8}{16}$ form a proportion.

   How do you know if the fractions (ratios) form a proportion?

   OPTION 1:

   Simplify each fraction

   $\frac{2}{4} = \frac{1}{2}$ and $\frac{8}{16} = \frac{1}{2}$

   Both fractions simplify to $\frac{1}{2}$ so $\frac{2}{4}$ *and* $\frac{8}{16}$ form a proportion.

   OPTION 2:

   Use the cross products

   $\frac{2}{4} \times \frac{8}{16}$ $\quad 4 \cdot 8 = 32$, $\quad 2 \cdot 16 = 32$

   Since the cross products are equal, $\frac{2}{4}$ *and* $\frac{8}{16}$ form a proportion.

2. A recipes calls for 2 cups of flour and makes 20 cookies. If you want to make 35 cookies, how many cups of flour will you need?

"How many cups of flour will you need" is the unknown. Assign this a variable.

What do we know?

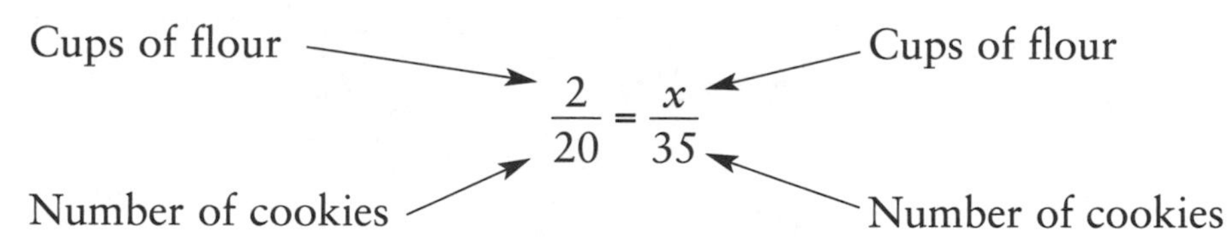

Find the cross products. $2 \cdot 35 = 20 \cdot x$

Multiply. $\frac{70}{20} = \frac{20x}{20}$

Divide each side by 20.

Simplify. $3.5 = x$

Solution: You will need 3.5 cups of flour to make 35 cookies.

TEST YOUR SKILLS (For answers, see page 255.)

1. Your grandmother's favorite cookie recipe calls for 2 cups of flour to make 12 cookies. If your grandmother wants to make 60 cookies, how many cups of flour will she need?

A. 10 cups
B. 12 cups
C. 15 cups
D. 20 cups

2. Joe needs to put gas in his car. If 10 gallons of gas cost him $25, how much will it cost Joe to buy 12 gallons of gas?

A. $27
B. $30
C. $48
D. $49

3. Kelly and Tracy love to text message. In the amount of time it takes Kelly to text 12 words, Tracy can text 15 words. If Kelly can text message 20 words, how many words will Tracy text in that time?

A. 12

B. 16

C. 23

D. 25

4. In a recent survey, three out of eight households watch "I Want to Be an Idol." If 45 households said they watched "I Want to Be an Idol," how many households were surveyed?

A. 17

B. 60

C. 120

D. 135

5. Your parents have decided to pay you an allowance based on the number of loads of laundry that you do. They said that for every three loads of laundry you wash and dry, they will pay you \$1.50. If you wash and dry five loads of laundry, how much money will your parents owe you?

A. \$2.50

B. \$3.00

C. \$10.00

D. \$25.00

6. Kalien likes to download songs from the Internet. She can download 10 songs for \$2.00. How much will it cost Kalien to download 25 songs?

A. \$2.50

B. \$5.00

C. \$7.50

D. \$8.00

7. Dylan has joined the school track team. He can run 20 laps around the track in 5 minutes. At this rate, how many laps will Dylan be able to run around the track in 20 minutes?

A. 80

B. 85

C. 90

D. 100

Short Response

8. You are having your friends over for a pizza party. If it costs $15 for 24 pieces of pizza, how much money will you spend if you need 36 pieces of pizza?

Show your work.

Answer __________

9. Determine if the pair of ratios form a proportion:

$$\frac{6}{18} \text{ and } \frac{24}{72}$$

Part A.

Show your work.

Answer __________

Part B.

Explain one method of how you can determine whether two ratios form a proportion.

____

____

____

SOLVING AND GRAPHING 1-STEP INEQUALITIES

INEQUALITIES

An inequality is a mathematical sentence that uses >, <, ≥, or ≤.

Let's review what the symbols stand for.

Symbol <	Symbol >
is less than is fewer than	is greater than is more than exceeds
Symbol ≤	**Symbol ≥**
is less than or equal to is no more than is at most	is greater than or equal to is at least is no less than

Inequalities represent many situations or phrases in your life that you may not even be aware of.

Examples

Phrase	Inequality
You must be 18 years or older to vote.	$a \geq 18$
Children under the age of 6 eat free.	$c < 6$
You must be at least 48 inches tall to ride the rollercoaster.	$h \geq 48$
Children over the age of 15 may attend the dance.	$a > 15$

To solve an inequality, you use the same steps you would use to solve an equation. The important concept to remember is that when you solve an equation, you get one answer. When you solve an inequality, there are many answers.

Equation	Inequality
$2x + 5 = 11$	$2x + 5 > 11$
$-5 \quad -5$	$-5 \quad -5$
$\frac{2x}{2} = \frac{6}{2}$	$\frac{2x}{2} > \frac{6}{2}$
$x = 3$	$x > 3$
$x = 3$ means the only answer that will work in the equation is 3.	$x > 3$ means that any number greater than 3 will work in this inequality.
	Example: 3.1, 4, 10, and 20 are just a few numbers that are solutions for the inequality.

Since inequalities have more than one solution, we graph inequalities on a number line to show the solution set.

You need to remember:

	Phrase	To graph use:
$>$	is greater than	open circle ○→
$<$	is less than	open circle ←○
$\geq$	is greater than or equal to	closed circle ●→
$\leq$	is less than or equal to	closed circle ←●

> HINT! When reading or graphing inequalities, always make sure the letter comes first!
> Example: $3 > x \rightarrow$ is the same as $x < 3$.
> This is easier to read and graph.

Examples

Graph the solution set.

1. $x > 5$

2. $x \leq -2$

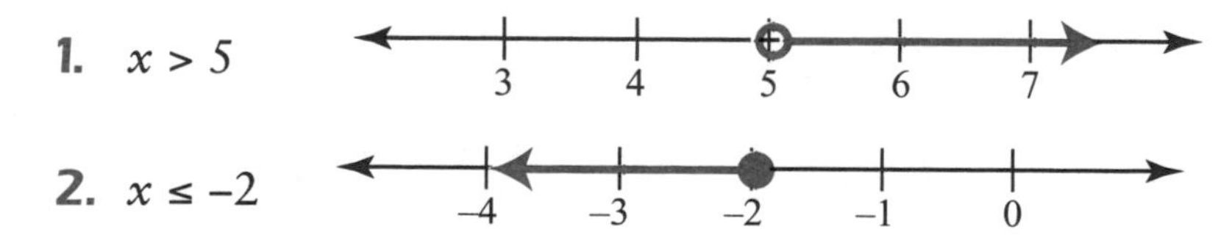

3. $x \geq -7$

4. $x < 4$

Solve and graph the following inequalities.

5. $x + 18 > 26$

$$\begin{array}{rr} x + 18 > & 26 \\ -18 & -18 \\ \hline x & > 8 \end{array}$$

6. $x - 10 \leq 3$

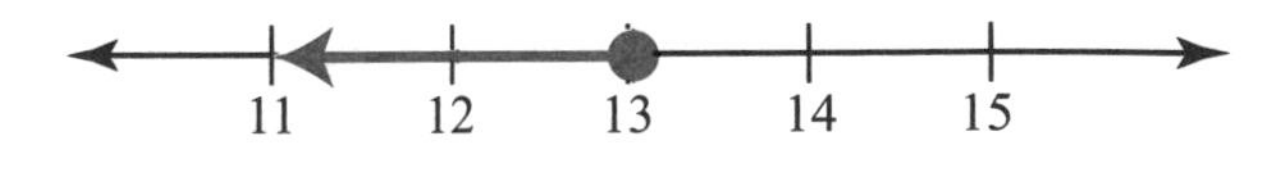

$$\begin{array}{rr} x - 10 \leq & 3 \\ +10 & +10 \\ \hline x & \leq 13 \end{array}$$

7. $y - 8 \geq 1$

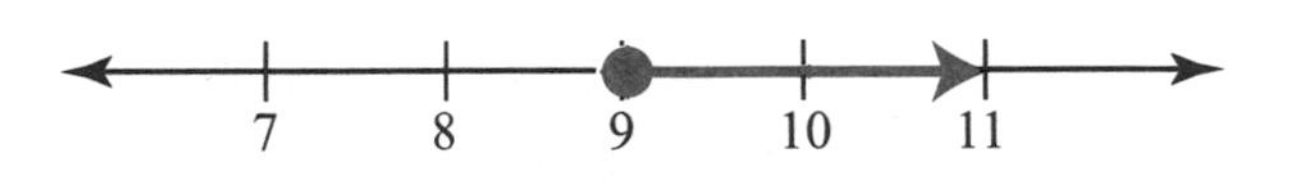

$$\begin{array}{rr} y - 8 \geq & 1 \\ +8 & +8 \\ \hline y & \geq 9 \end{array}$$

8. $5 > c + 4$

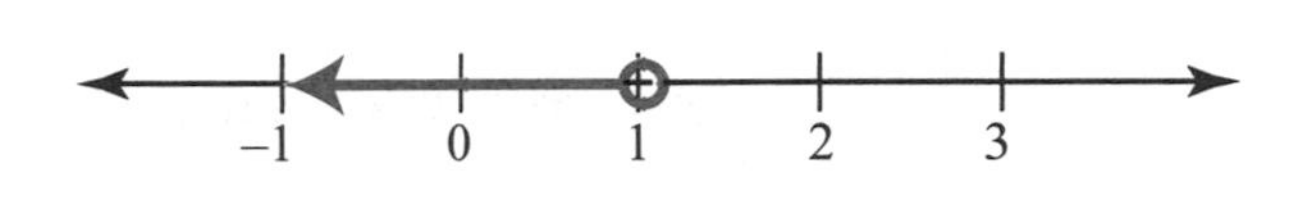

$$\begin{array}{rr} 5 > & c + 4 \\ -4 & -4 \\ \hline 1 & > c \end{array}$$

(which reads $c < 1$)

TEST YOUR SKILLS (For answers, see page 256.)

1. Solve: $x + 5 \geq 12$

A. $x \leq 7$

B. $x \geq 7$

C. $x \geq 17$

D. $x \leq 17$

2. Solve: $y - 7 < 9$

A. $y \leq 2$

B. $y \geq 2$

C. $y < 16$

D. $y > 16$

3. Solve: $\frac{m}{2} > 5$

A. $m > 3$

B. $m > 7$

C. $m < 10$

D. $m > 10$

4. Which number is *not* included in the solution set for the inequality $2d \leq 20$?

A. 0

B. 5

C. 10

D. 11

5. Which phrase below represents the solution set of the inequality $4 + y > 5$?

A. y is less than 5

B. y is greater than 5

C. y is greater than 9

D. y is less than or equal to 9

6. Which graph shows the solution for the inequality $x - 7 \leq 9$?

A.

B.

C.

D.

7. Which graph shows the solution for the inequality $3 + y > 7$?

A. 

B. 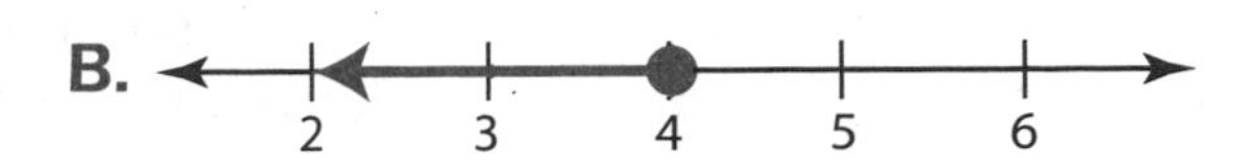

C. 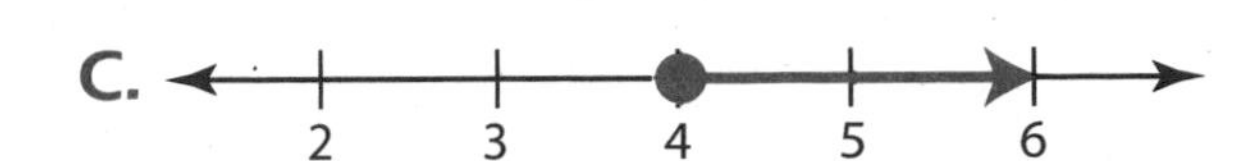

(D.) 

8. The number line below is the graph of which inequality?

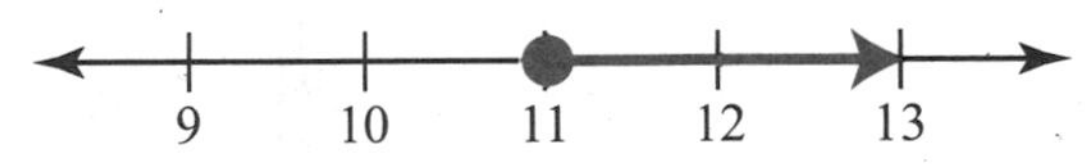

A. $x + 3 > 8$

B. $x - 3 > 8$

C. $x + 3 \le 8$

(D.) $x - 3 \ge 8$

Short Response

9. Solve the inequality: $3 \ge y - 4$

Part A.

Show your work.

$3 \ge y - 4$
$+4 \quad +4$
$7 \ge y$

Part B.

Graph the solution set for the inequality in Part A.

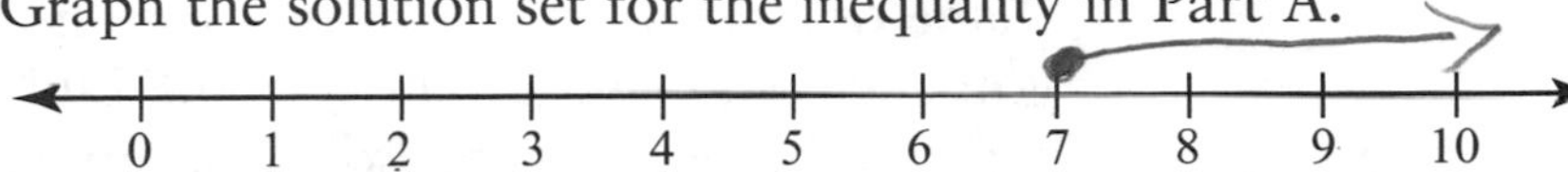

EVALUATING FORMULAS USING SURFACE AREA, RATE, AND DENSITY

A formula is an equation that shows a mathematical relationship between two or more quantities. We use formulas to help us solve many different problems. In this section, you will be given a formula and some information about the formula. Your goal is to plug in the information you know to find the answer.

Surface Area

Cube

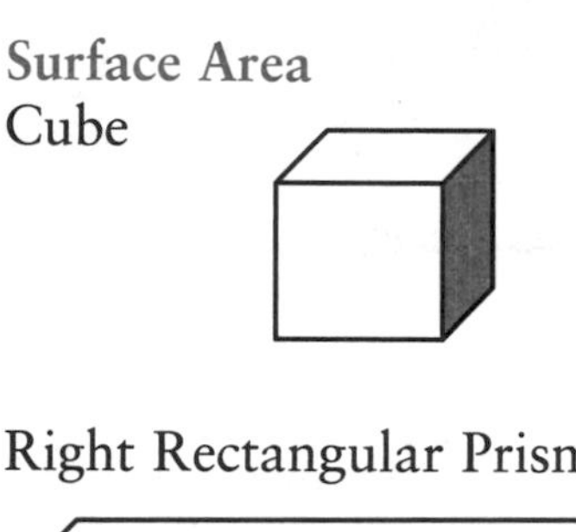

Total Surface Area = $6s^2$

Right Rectangular Prism

Total Surface Area = $2wl + 2lh + 2wh$

Rate $I = prt$

i = amount of interest
p = principal (amount invested – started with)
r = rate of interest (in decimal form)
t = time (in years)

Density $d = m \div v$

d = density
m = mass
v = volume

Examples

1. You need to wrap a gift for your friend's birthday. The gift is the shape of a rectangular prism. To see how much wrapping paper you need, you can use the formula $2wl + 2lh + 2wh$ to find the total surface area of the box. You know the length of the box is 6 inches, the width is 4 inches, and the height is 3 inches. Find the total surface area of the box.

Solution: You know the length of the box is 6 inches, the width is 4 inches and the height is 3 inches. Substitute these values into the formula.

Rewrite the formula.	$2wl + 2lh + 2wh$
Substitute the values.	$2(4)(6) + 2(6)(3) + 2(4)(3)$
Follow the order of operations.	$48 + 36 + 24$
Add to find the total.	108 square inches

2. The formula for converting temperature from degrees Celsius to degrees Fahrenheit is $F = \frac{9}{5}C + 32$. What is the temperature in degrees Fahrenheit when the temperature is 10° Celsius?

Solution: Substitute 10 in the formula for C.

Rewrite the formula.	$F = \frac{9}{5}C + 32$
Substitute 10 in for C.	$F = \frac{9}{5}(10) + 32$
Multiply $\frac{9}{5}(10)$.	$F = 18 + 32$
Add.	F = 50 degrees

TEST YOUR SKILLS (For answers, see page 257.)

1. For your birthday, your aunt has given you \$75. You decide to deposit this into your bank account. Find the simple interest if you keep the money in the bank for 5 years at a rate of 5.5%. Round your answer to the nearest cent.

A. \$2.63

B. \$20.63

C. \$206.25

D. \$2062.50

2. The formula for the surface area of a cube is $6s^2$. Find the surface area of a cube with a side length of 3 inches.

A. 36 sq. in.
B. 54 sq. in.
C. 81 sq. in.
D. 324 sq. in.

3. The volume of a right triangular prism is $\frac{1}{2}lwh$, where w is the width, h is the height, and l is the length. Find the volume of a right triangular prism that has a width of 4 inches, a height of 5 inches, and a length of 8 inches.

A. 80 cu. in.
B. 160 cu. in.
C. 200 cu. in.
D. 425 cu. in

4. The formula for finding the density of an object is $d = m \div v$; where d stands for density, m stands for mass, and v stands for volume. Find the density of an object with a mass of 725 kilograms and a volume of 25 cubic meters.

A. 29 kg/m^3
B. 30 kg/m^3
C. 200 kg/m^3
D. 18,125 kg/m^3

5. Suppose you invest $1000 in a CD that earns 3.5% per year. How much simple interest would your money earn after 2 years? Use the formula I = prt.

A. $7
B. $70
C. $700
D. $7000

6. The formula for finding distance is *distance* = *rate* × *time*. Find the distance traveled if you traveled 50 miles per hour for 7.5 hours.

   A. .15 miles

   B. 7 miles

   C. 37 miles

   D. 375 miles

7. The formula for finding the density of an object is $d = m \div v$; where d stands for density, m stands for mass, and v stands for volume. Find the mass of an object with a density of 60 kg/m^3 and a volume of 15 cubic meters.

   A. 4 kg

   B. 825 kg

   C. 900 kg

   D. 1000 kg

Short Response

8. Suppose you deposit $500 into a savings account with a simple interest rate of 6.5%.

Part A.

How much interest will the account earn in 4 years? Use the formula I = prt. Show your work.

Answer __________

Part B.

Find the total amount in the savings account after 4 years. Show your work.

Answer __________

Chapter 3

GEOMETRY

PLOTTING POINTS ON THE COORDINATE PLANE

The coordinate plane is formed by the intersection of the horizontal and vertical number lines.

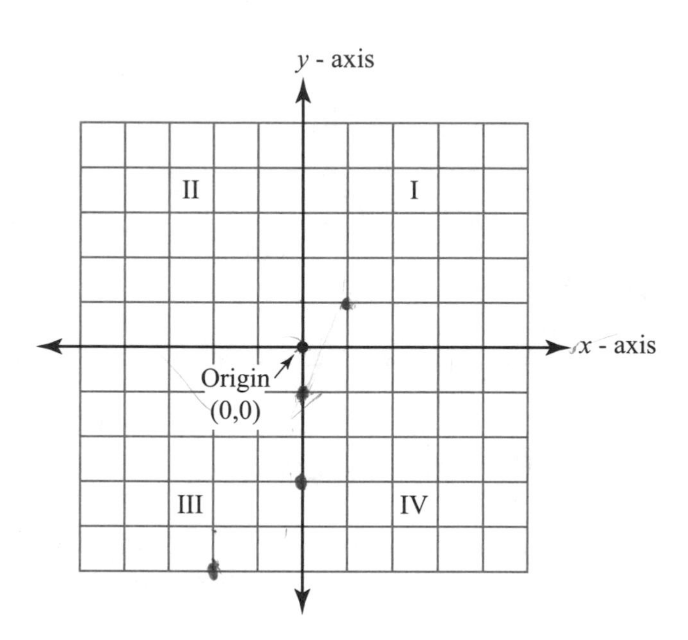

The lines with the arrows are called the axes.

The x-axis is the horizontal line that goes left and right.
The y-axis is the vertical line that goes up and down. (To help you remember, use the saying "Y in the sky.")
The origin is the point where the x- and y-axis cross. The coordinate of the origin is (0, 0).
An ordered pair are the points (x, y) in the coordinate plane.

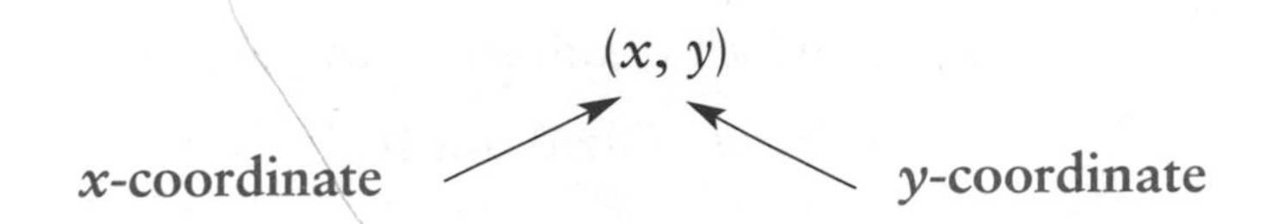

Positive x moves to the right.
Negative x moves to the left.

Positive y moves up.
Negative y moves down.

To graph the ordered pair (x, y), start at the origin (0, 0) and move left or right, and then up or down.

> The **quadrants** are the four areas on the coordinate plane formed by the x- and y-axes.

Examples:

Explain the direction of each ordered pair from the origin.

(3, 5)	3 units to the right, 5 units up	(0, 3)	0 units left or right, up 3 units
(–2, 6)	2 units to the left, 6 units up	(–4, 0)	4 units to the left, 0 units up or down
(–5, –9)	5 units to the left, 9 units down		

Examples

1. Graph the point A (5, –2).

Solution:

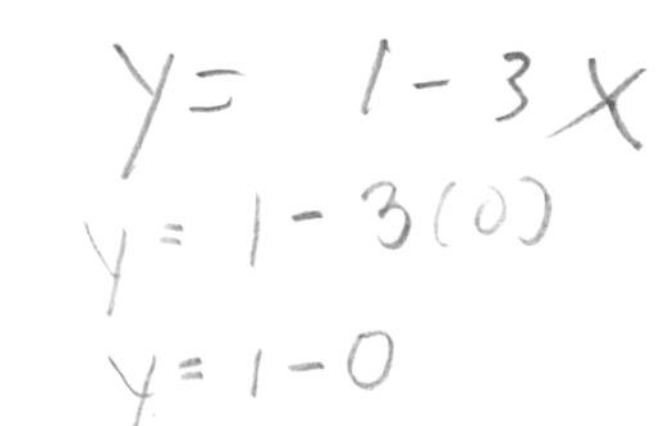

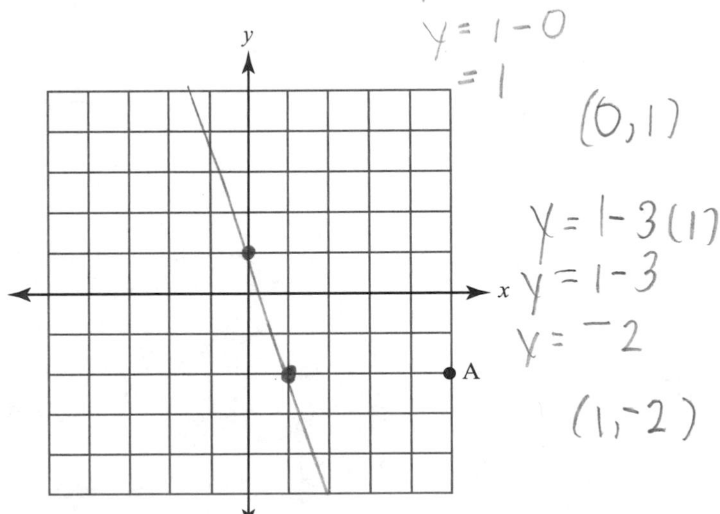

2. In which quadrant does the point (–2, 6) lie?

Solution: Quadrant II

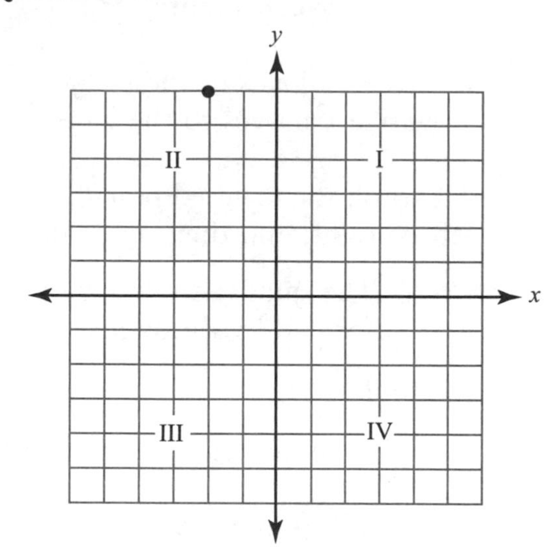

TEST YOUR SKILLS (For answers, see page 259.)

1. Which letter identifies the ordered pair (–3, –6)?

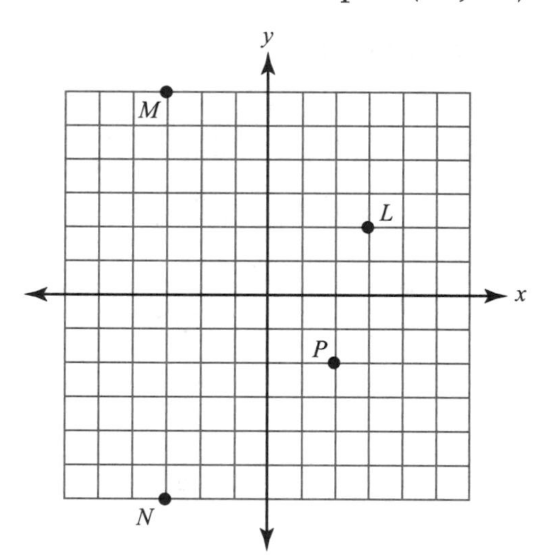

A. L

B. M

C. N

D. P

2. Which letter identifies the ordered pair (0, 3)?

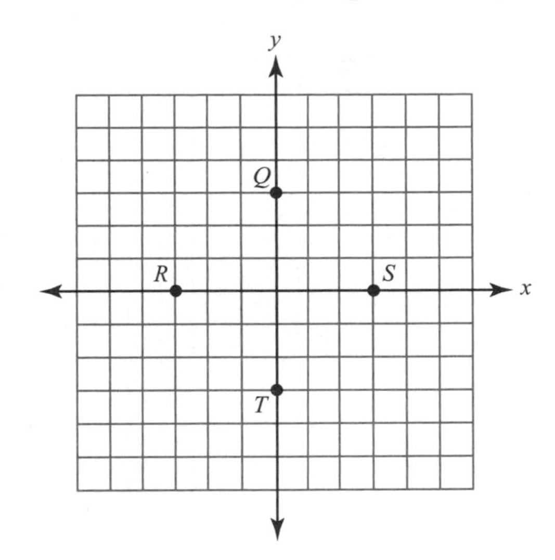

A. *Q*

B. *R*

C. *S*

D. *T*

3. Which four ordered pairs are shown on the graph below?

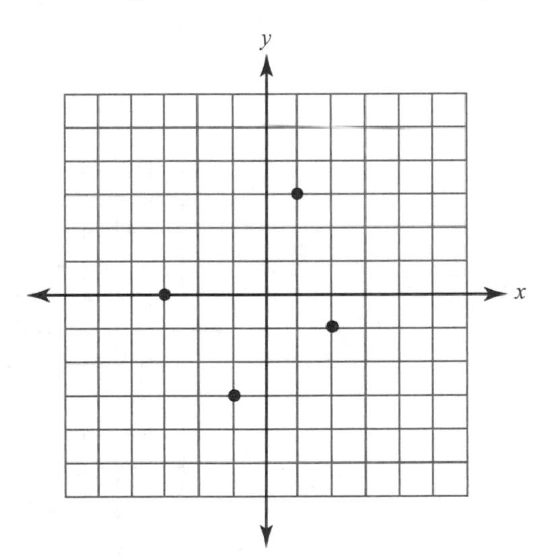

A. (–3, 0), (–1, 3), (0, 1), (6, 4)

B. (–3, 0), (–1, –3), (1, 3), (2, –1)

C. (0, –3), (3, –1), (1, 3), (4, 2)

D. (0, –3), (1, –3), (2, 1), (4, 2)

4. Which three ordered pairs are shown on the graph below?

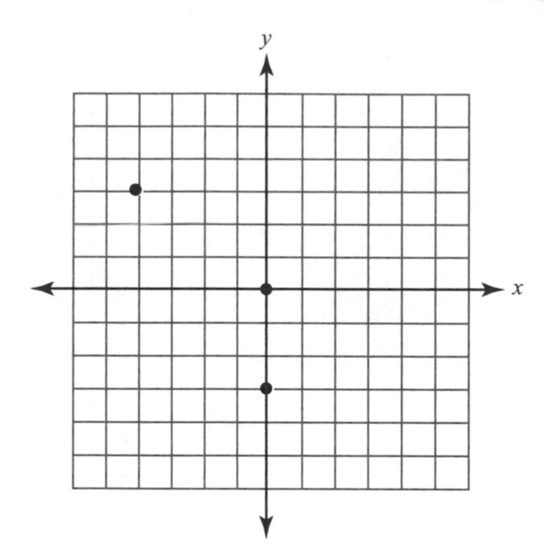

A. (4, 3), (2, 3), (1, 1)

B. (0, 0), (3, 0), (4, 3)

C. (3, –4), (2, 1), (1, 4)

D. (–4, 3), (0, 0), (0, –3)

5. Which ordered pair will complete the figure so that the shape will be a rectangle?

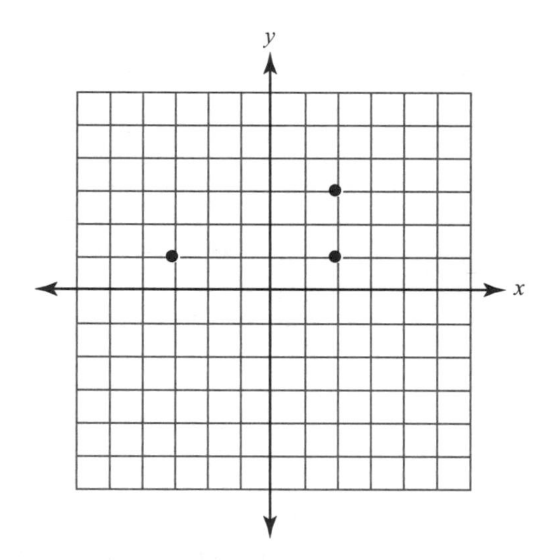

A. (–3, –3)

B. (4, 1)

C. (–3, 3)

D. (–2, 2)

6. Which ordered pair is *not* shown on the graph below?

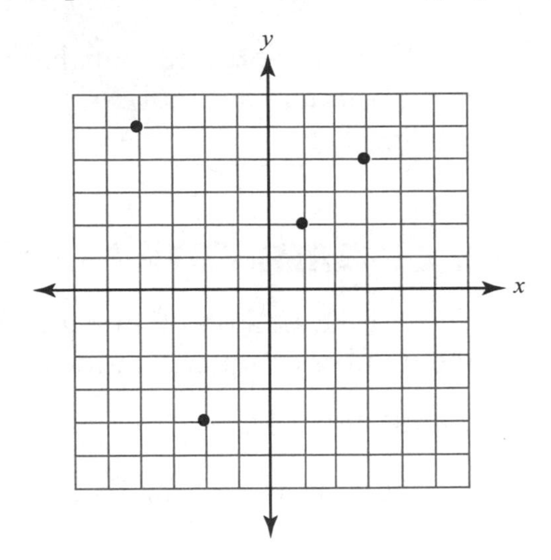

A. (–2, –4)

B. (3, 4)

C. (–4, 5)

D. (4, –2)

Short Response

7. **Part A.**

Graph and label the three points on the coordinate graph below.

$A(-3, 3)$, $B(2, 3)$, $C(-3, -3)$

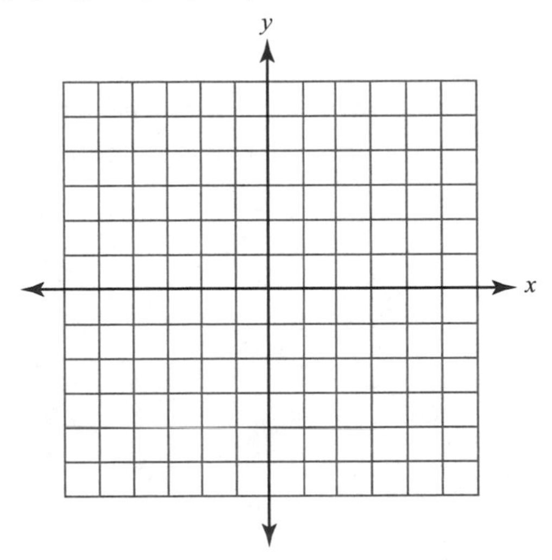

Part B.

Name the ordered pair that will complete the figure so the shape is a rectangle.

Answer __________

AREA ON THE COORDINATE PLANE

To find the area of a figure graphed on a coordinate plane, count the number of squares inside the figure.

Examples

1. Find the area of the figure.

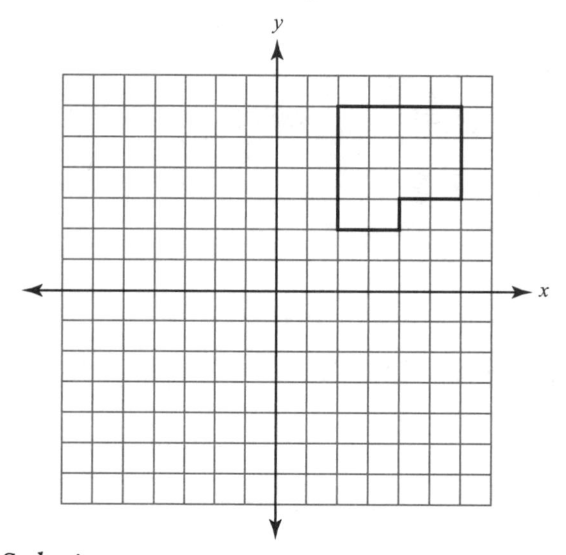

Solution:

Count the number of square units inside the figure. There are 14 square units inside the figure.

2. How many figures can the figure be divided into?

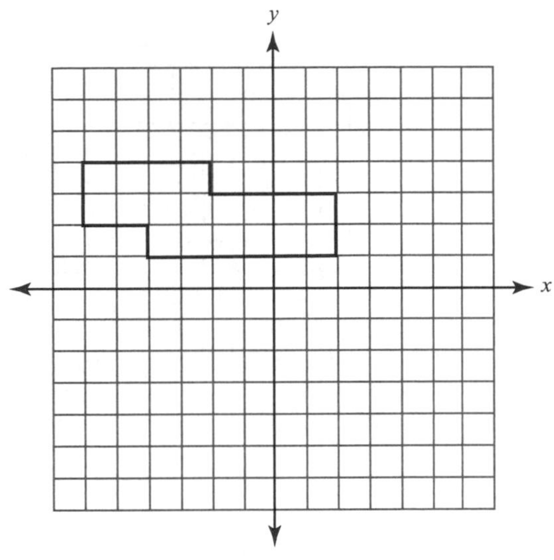

Solution:

The figure can be divided into 1 square and 2 rectangles.

TEST YOUR SKILLS (For answers, see page 260.)

1. What is the area of the playground shown on the graph below?

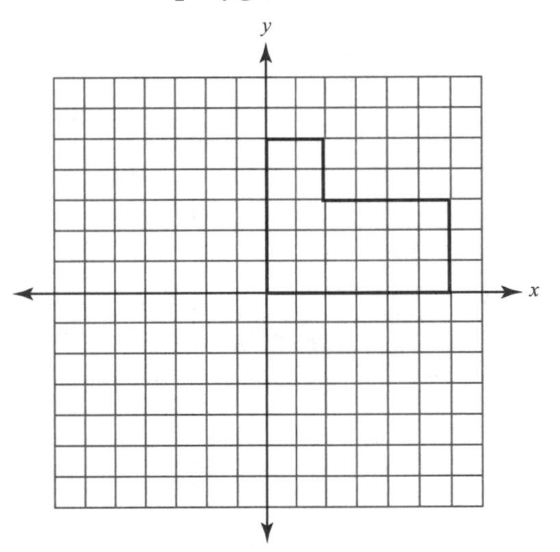

A. 18 square units
B. 22 square units
C. 24 square units
D. 30 square units

2. What is the area of the figure given below?

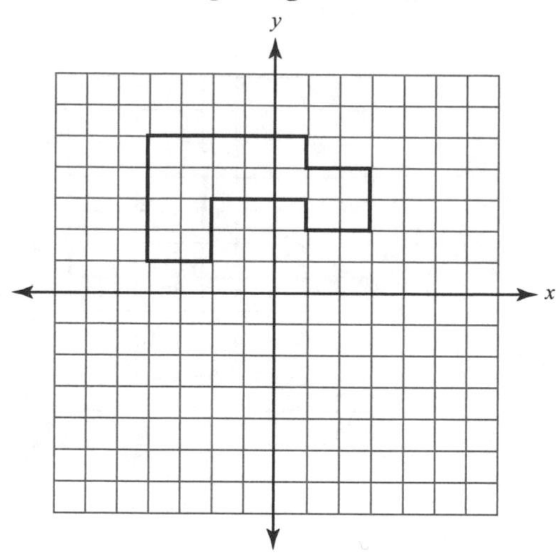

A. 10 square units

B. 18 square units

C. 23 square units

D. 28 square units

3. What three figures can you divide the figure into?

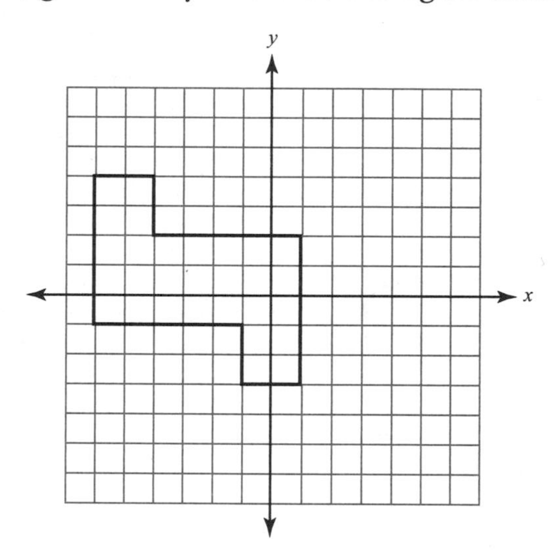

A. 3 squares

B. 3 triangles

C. 3 rectangles

D. 2 rectangles, 1 square

4. What figures can you divide the figure into?

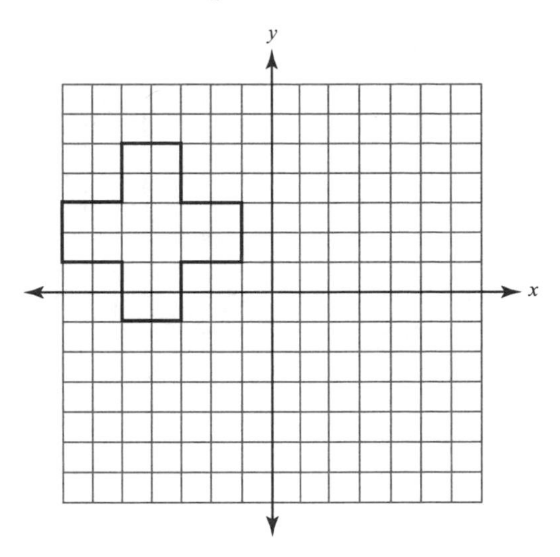

A. 4 squares

B. 5 squares

C. 2 rectangles

D. 3 triangles

5. Which figure has an area that is different than the areas of the other three figures?

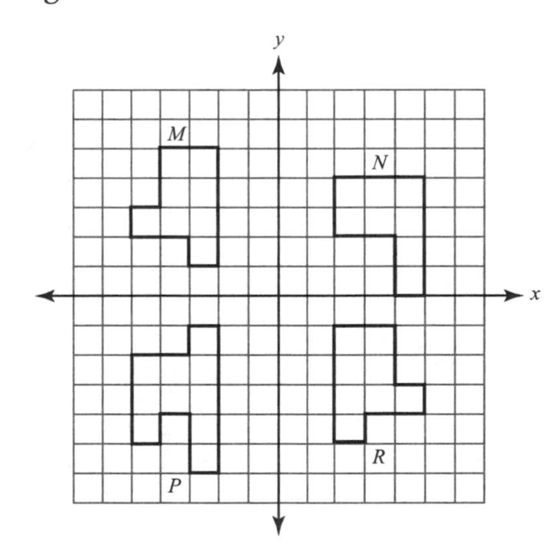

A. *M*

B. *N*

C. *P*

D. *R*

6. Which figure has an area of 7 square units?

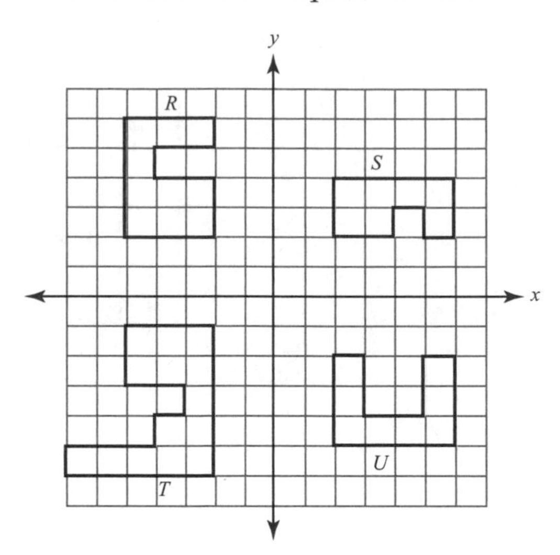

A. *R*

B. *S*

C. *T*

D. *U*

Short Response

7. On the graph below, draw a figure with an area of 42 square units.

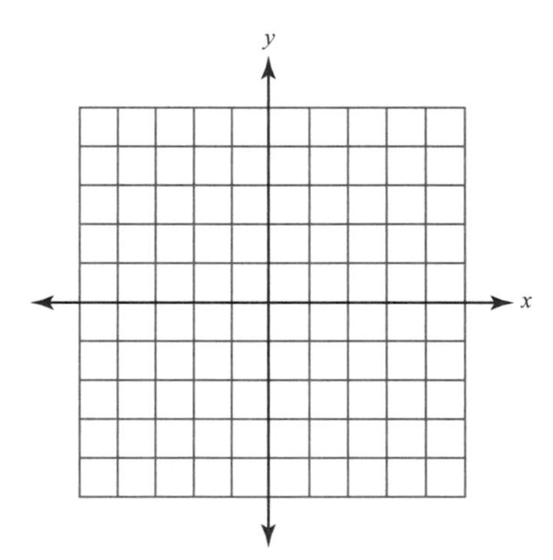

AREA AND CIRCUMFERENCE

PARTS OF A CIRCLE

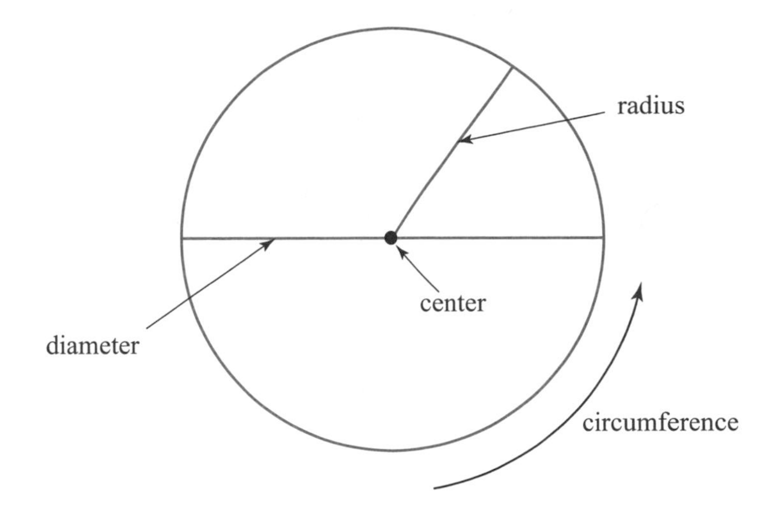

Radius	A segment connecting the center of the circle and a point on the circle. The radius is half of the diameter.
Diameter	A segment connecting two points on the circle and passing through the center of the circle.
Circumference	The distance around the circle.
Area	The number of square units in a circle.

CIRCLE FORMULAS

Diameter	$d = 2r$	use to find diameter or the radius
Circumference	$C = \pi d$	use to find circumference or the diameter
Circumference	$C = 2\pi r$	use to find circumference or the radius
Area	$A = \pi r^2$	use to find area or the radius

Using this information, you can solve for thc diameter, radius, area, or circumference of a circle.

Examples

1. Find the diameter of a circle if the circumference of the circle is 78.5 meters. Use 3.14 for π.

Solution:	Since you want to find the diameter, use $C = \pi d$	$C = \pi d$
	Substitute 78.5 for C	$78.5 = \pi d$
	Let $\pi = 3.14$	$78.5 = 3.14d$
	Divide by 3.14	$d = 25$

The diameter equals 25 meters.

2. What is the radius of a circle whose area is 50.24 square meters? Use 3.14 for π.

Solution:	Use the formula $A = \pi r^2$	$A = \pi r^2$
	Substitute 50.24 for A and 3.14 for π	$50.24 = 3.14r^2$
	Divide by 3.14	$16 = r^2$
	Take the square root of r and 16	$4 = r$

The radius of the circle is equal to 4 meters.

3. Find the circumference of a circle with an area equal to 25π square inches. Leave your answer in terms of π.

Solution:	Since you are given the area of the circle, use the area formula $A = \pi r^2$	$A = \pi r^2$
	Substitute 25π in for A	$25\pi = \pi r^2$
	Solve for r and divide both sides by π	$25 = r^2$
	Take the square root of r and 25	$5 = r$
	Now that you know $r = 5$, find the circumference	$C = 2\pi r$
	Substitute 5 in for r. Leave π	$C = 2(\pi)(5)$
	Multiply	$C = 10\pi$

The circumference is 10π inches.

4. Find the area of a circle with a circumference of 16π meters. Leave your answer in terms of π.

Solution: Since you are given the circumference of the circle,

use the formula $C = 2\pi r$	$C = 2\pi r$
Substitute 16π in for C	$16\pi = 2\pi r$
Divide by 2π	$8 = r$
Now that you know $r = 8$, find the area of the circle	$A = \pi r^2$
Substitute 8 in for r, leave in terms of π	$A = \pi(8^2)$
Simplify	$A = 64\pi$

The area of the circle is 64π square meters.

TEST YOUR SKILLS (For answers, see page 260.)

1. Find the diameter of a circle with a circumference of 62.8 inches. Use 3.14 for π.

A. 10 inches

B. 20 inches

C. 60 inches

D. 197 inches

2. Find the radius of a circle if the area is 28.26 square inches. Use 3.14 for π.

A. 3 inches

B. 4.5 inches

C. 5 inches

D. 7 inches

3. Find the circumference of a circle with an area of 49π square meters. Leave your answer in terms of π.

A. 7π m

B. 12π m

C. 14π m

D. 21π m

4. Find the area of a circle with a circumference of 60π inches. Leave your answer in terms of π.

A. 30π in^2

B. 60π in^2

C. 600π in^2

D. 900π in^2

5. Find the radius of a circle if the circumference is 12π centimeters.

A. 4 cm

B. 6 cm

C. 8 cm

D. 12 cm

6. What is the area of a circle if the circumference is 36π centimeters? Leave your answer in terms of π.

A. 18π cm^2

B. 36π cm^2

C. 324π cm^2

D. 1296π cm^2

7. One circle has a radius of 5. A second circle has a radius of 6. How much larger is the circumference of the second circle than the first circle?

A. 2π

B. 3π

C. 4π

D. 5π

Short Response

8. Joanna and Steve have been given two different circles, a red circle and a blue circle. The red circle has an area of 16π and the blue circle has a circumference of 12π.

   Which circle has the larger radius? Show your work.

   Answer __________

MISSING ANGLES IN QUADRILATERALS

A quadrilateral is a two-dimensional geometric figure with four sides.

The sum of the angles in any quadrilateral is 360°.

Recall that the sum of the angles in any triangle is 180°.

Example

Find the value of the missing angle.

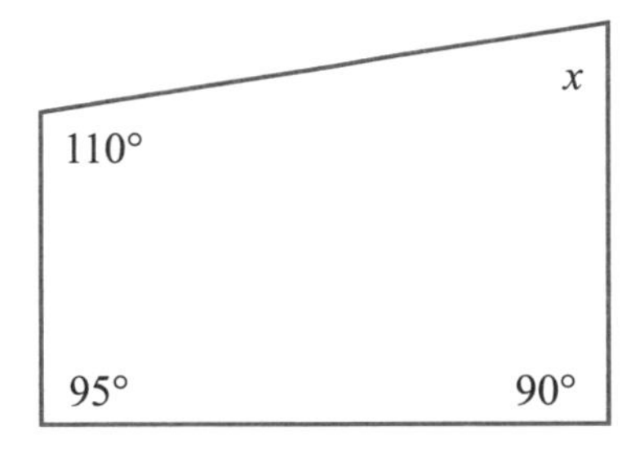

Solution:

The angles in a quadrilateral add up to 360°.

$110 + 95 + 90 + x = 360$

$$\begin{array}{r} 295 + x = 360 \\ \underline{-295 \qquad -295} \\ x = 65 \end{array}$$

The missing angle is 65°.

TEST YOUR SKILLS (For answers, see page 261.)

1. What is the measure of the missing angle in the quadrilateral?

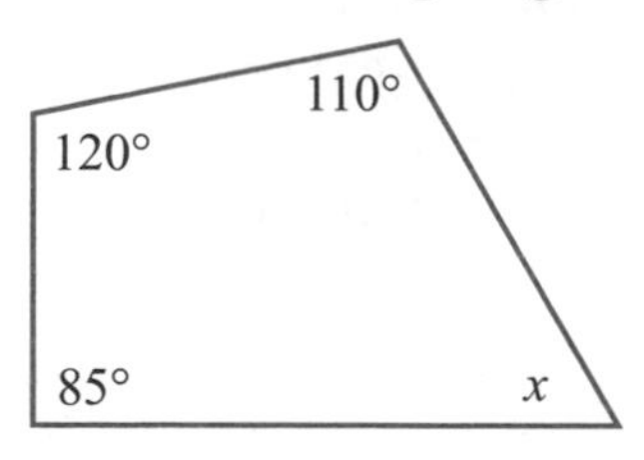

A. 45°

B. 55°

C. 100°

D. 135°

2. Two angles in a quadrilateral are 65° and 95°. The other angles are equal. What is the measure of each of the other angles?

A. 80°

B. 90°

C. 100°

D. 110°

3. One angle in a quadrilateral measures 120°. What are the possible measures of the other three angles?

A. 40°, 60°, 60°

B. 60°, 80°, 120°

C. 70°, 70°, 70°

D. 70°, 80°, 90°

4. The sum of two angles in a quadrilateral is 140°. Find the measures of the other two angles if the angles are equal.

A. 55°

B. 80°

C. 110°

D. 220°

5. Which angle measures could be the angles in a quadrilateral?

   A. 30°, 60°, 90°, 100°
   B. 60°, 60°, 70°, 150°
   C. 70°, 60°, 90°, 120°
   D. 80°, 80°, 150°, 50°

6. One angle in a quadrilateral is 150°. What are the possible measurements of the other three angles?

   A. 30°, 70°, 110°
   B. 40°, 50°, 60°
   C. 45°, 45°, 60°
   D. 120°, 90°, 35°

7. Which angle measurements could *not* be the angles in a quadrilateral?

   A. 70°, 70°, 86°, 134°
   B. 90°, 90°, 45°, 100°
   C. 90°, 90°, 90°, 90°
   D. 100°, 60°, 50°, 150°

Short Response

Use what you know about the sum of angles in quadrilaterals to answer the following questions.

8. **Part A.**

   List four possible angle measures of a quadrilateral if none of the angles are equal.

   Show your work.

   Answer __________

Part B.

List four possible angle measures of a quadrilateral if two of the angles are equal. Show your work.

Answer __________

Part C.

Is it possible for a quadrilateral to have four equal angles? If not, explain why. If yes, give the measures of the angles. Show your work.

Answer __________

FACES AND BASES OF 3D SHAPES

A three-dimensional figure is also known as a solid.

In this section, we will review the shapes that make up the faces and bases of prisms, cylinders, cones, and pyramids.

Vocabulary Review

Faces	The flat surface of the three-dimensional figure
Edges	The line segments where the faces meet
Bases	The parallel faces
Vertices	The points at the corners of the solid; where the edges meet

TYPES OF SOLIDS

Prisms

Rectangular Prism

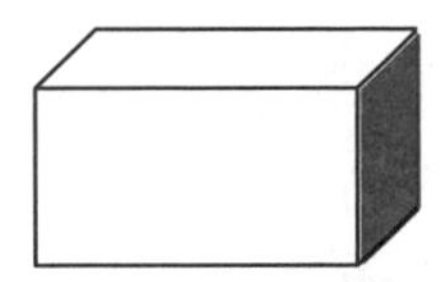

6 faces, all rectangles
12 edges
8 vertices

Cube

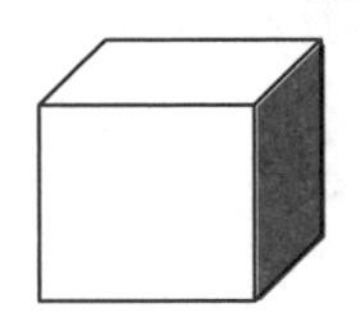

6 faces, all squares
12 edges
8 vertices

Triangular Prism

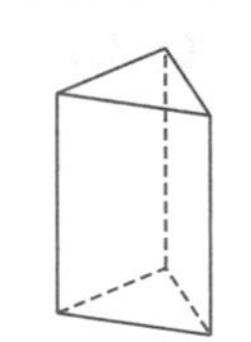

5 faces; the 2 bases are triangles, the 3 faces are rectangles
9 edges
6 vertices

Pyramids

Rectangular Pyramid

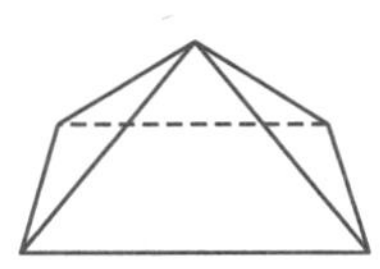

5 faces; the base is a rectangle, 4 faces are triangles
8 edges
5 vertices

Square Pyramid

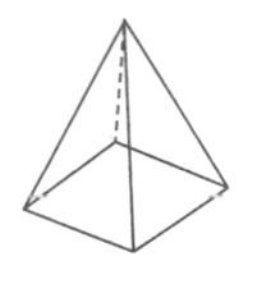

5 faces; the base is a square
8 edges
5 vertices

Triangular Pyramid

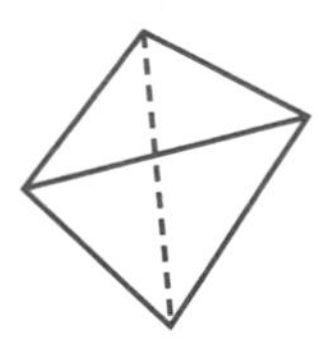

4 faces; all faces and its base are triangles
6 edges
4 vertices

Other

Cylinder

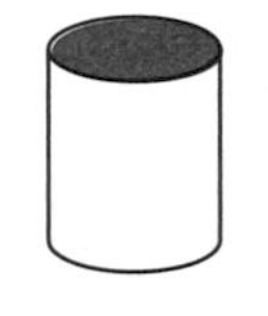

2 faces; the faces are circles
no edges
no vertices

Cone

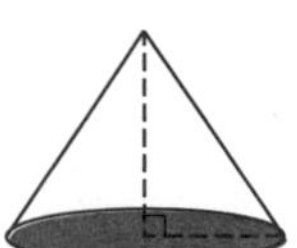

1 face; the face is a circle
no edges
1 vertex

Examples

Use the figure below to answer questions 1–3.

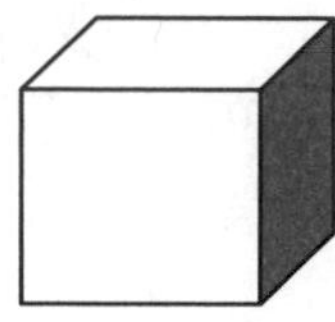

1. How many faces does the figure have?

   Solution: The figure has 6 faces (front, back, 2 sides, top, and bottom).

2. How many edges does the figure have?

   Solution: The figure has 12 edges.

3. How many vertices does the figure have?

   Solution: The figure has 8 vertices.

TEST YOUR SKILLS (For answers, see page 262.)

Use the figure shown below to answer questions 1–3.

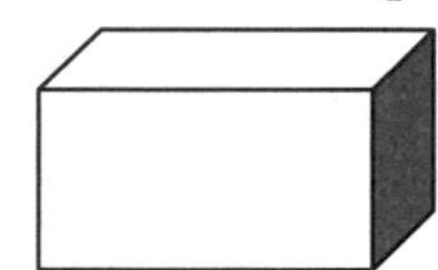

1. How many faces does the figure have?

   A. 5
   B. 6
   C. 8
   D. 10

2. How many vertices does the figure have?

   A. 8
   B. 10
   C. 12
   D. 14

3. What is the name of the solid figure?

   A. triangular prism
   B. rectangular prism
   C. triangular pyramid
   D. rectangular pyramid

Use the figure shown below to answer questions 4–7.

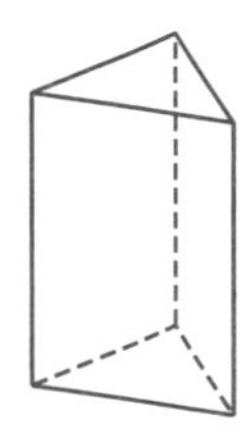

4. How many faces does the figure have?

   A. 4
   B. 5
   C. 6
   D. 8

5. How many bases does the figure have?

   A. 1
   B. 2
   C. 4
   D. 5

6. How many edges does the figure have?

   A. 9
   B. 10
   C. 12
   D. 14

7. What is the name of the solid figure?

A. triangular prism
B. triangular pyramid
C. rectangular pyramid
D. cylinder

Use the figure shown below to answer questions 8–10.

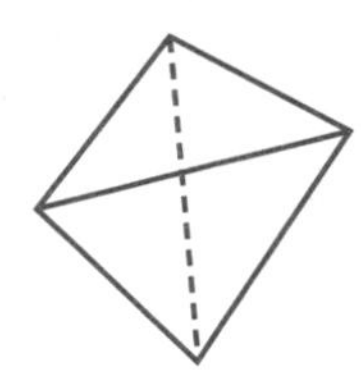

8. How many faces does the figure have?

A. 2
B. 3
C. 4
D. 6

9. How many vertices does the figure have?

A. 2
B. 3
C. 4
D. 5

10. What is the name of the solid figure?

A. cone
B. rectangular prism
C. triangular prism
D. triangular pyramid

Short Response

11. Identify the solid shown below. Name the number of faces, edges, and vertices.

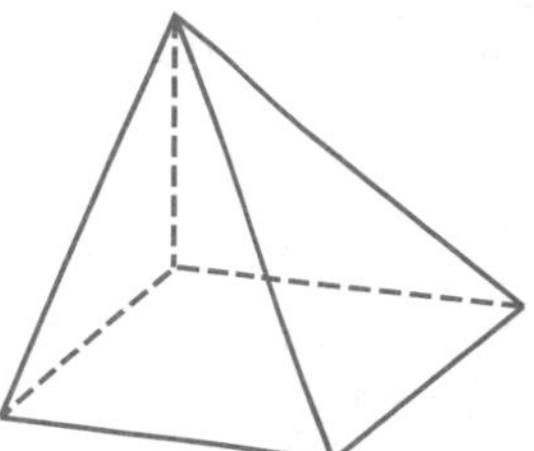

Name of solid ________

Number of faces ________

Number of edges ________

Number of vertices ________

VOLUME OF PRISMS AND CYLINDERS

The volume of a solid is the measure of the space inside of the solid. A unit for measuring volume is the cubic unit. Common measurements are cubic inches (in^3) or cubic feet (ft^3).

Volume of Rectangular Prisms

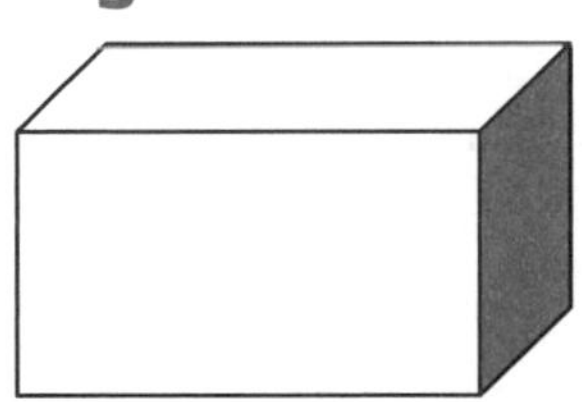

$V = lwh$

Where l = length
w = width
h = height

Volume of Cylinders

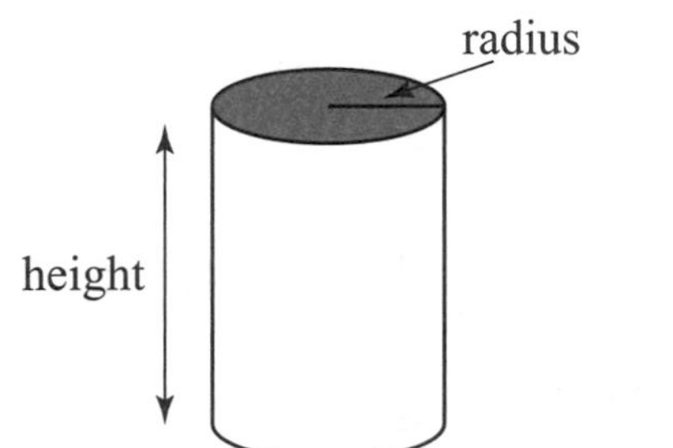

$V = \pi r^2 h$

Where r = radius
h = height

Examples

1. A toy box is the shape of a rectangular prism. Find the volume of the box if the length is 5 feet, the width is 3 feet, and the height is 3 feet.

Solution:

Use the formula $V = lwh$	$V = lwh$
Substitute $l = 5$, $w = 3$, $h = 3$	$V = 5(3)(3)$
Multiply	$V = 45$ cubic feet

The volume of the toy box is 45 cubic feet.

2. The swimming pool in your back yard has a 13 foot radius and is 5.5 feet high. How much water will the pool hold? Use 3.14 for π. Round your answer to the nearest tenth.

Solution:

Since the pool is a cylinder, use the formula $V = \pi r^2 h$	$V = \pi r^2 h$
Substitute $\pi = 3.14$, $r = 13$, $h = 5.5$	$V = 3.14(13^2)(5.5)$
Follow the order of operations	$V = 2918.63$
Round to the nearest tenth	$V = 2918.6$ cubic feet

The pool will hold 2918.6 cubic feet of water.

TEST YOUR SKILLS (For answers, see page 263.)

1. A cardboard box has a length of 5.5 feet, a width of 2.5 feet, and a height of 6 feet. What is the volume of the cardboard box?

A. 14 cubic feet

B. 82.5 cubic feet

C. 90 cubic feet

D. 108 cubic feet

2. What is the volume of a cylinder that has a radius of 3 inches and a height of 5 inches? Use 3.14 for π.

A. 47.1 cubic inches

B. 94.2 cubic inches

C. 141.3 cubic inches

D. 150.3 cubic inches

3. What is the height of a shoebox that has a volume of 432 cubic inches, a length of 8 inches, and a width of 6 inches?

   A. 6 inches

   B. 8 inches

   C. 9 inches

   D. 384 inches

4. The volume of a cylinder is 942 cubic inches. Find the height of the cylinder if the radius of the base of the cylinder is 5 inches. Use 3.14 for π.

   A. 12 inches

   B. 30 inches

   C. 60 inches

   D. 66 inches

5. What is the height of a cylinder if the volume of the cylinder is 301.44 cubic feet and the base of the cylinder has a diameter of 8 feet?

   A. 6 ft.

   B. 8 ft.

   C. 12 ft.

   D. 20 ft.

6. Find the width of a rectangular prism if the volume of the prism is 2340 cubic centimeters, the height is 13 centimeters, and the length is 15 centimeters.

   A. 10 cm

   B. 12 cm

   C. 15 cm

   D. 18 cm

7. Find the volume of a cylinder that has a height of 10 inches and a diameter of 6 inches. Leave your answer in terms of π.

A. 60π

B. 75π

C. 90π

D. 100π

8. Find the radius of an oatmeal can if the volume is 339.12 cubic centimeters and the height is 12 centimeters. Use 3.14 for π.

A. 3 cm

B. 5 cm

C. 6 cm

D. 10 cm

9. The volume of a rectangular prism is 630 cubic inches. What is one possible set of dimensions for the prism?

A. 6 in., 8 in., 10 in.

B. 7 in., 8 in., 9 in.

C. 7 in., 9 in., 10 in.

D. 9 in., 10 in., 10 in.

10. The volume of a cylinder is 75π cubic centimeters. What is one set of possible dimensions? Leave your answer in terms of π.

A. $r = 2$, $h = 4$

B. $r = 3$, $h = 3$

C. $r = 4$, $h = 5$

D. $r = 5$, $h = 3$

Short Response

11. Find the diameter of a cylinder if the volume is 1582.56 cubic feet and the height of the cylinder is 14 feet. Use 3.14 for π. Show your work.

Answer __________

SURFACE AREA OF PRISMS AND CYLINDERS

The surface area of a three-dimensional shape is the area of the outside surface of the shape. To find the total surface area, you would need to find the area of each side of the shape and add them all together.

A unit for measuring surface area is the square unit. Common measurements are square inches (in^2) or square feet (ft^2).

You will need to be able to find the surface area of the following four three-dimensional figures:

CUBE

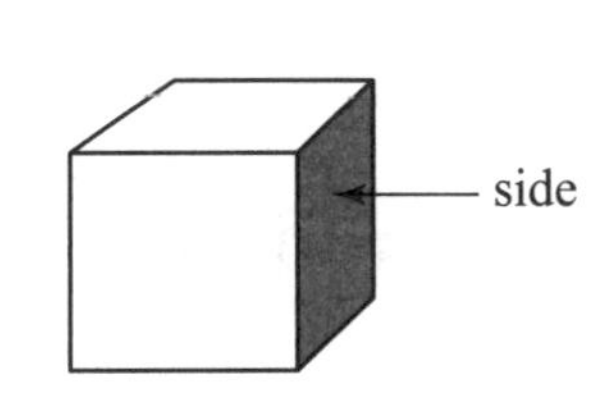

Total Surface Area = $6s^2$

RIGHT CIRCULAR CYLINDER

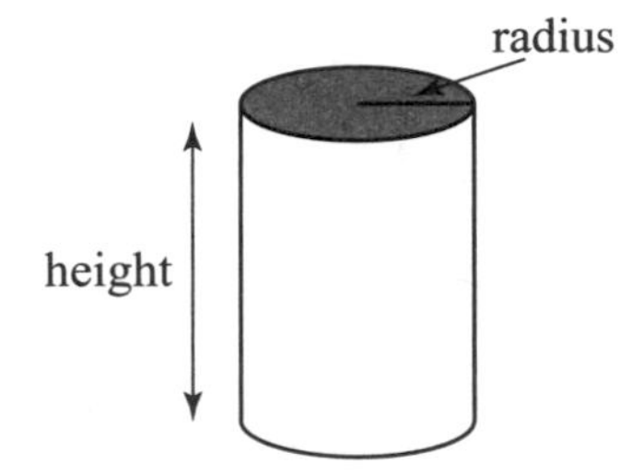

Total Surface Area = $2\pi rh + 2\pi r^2$

RIGHT RECTANGULAR PRISM

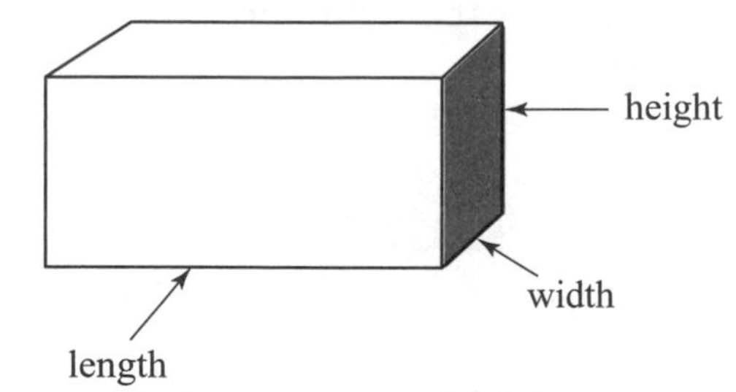

RIGHT TRIANGULAR PRISM

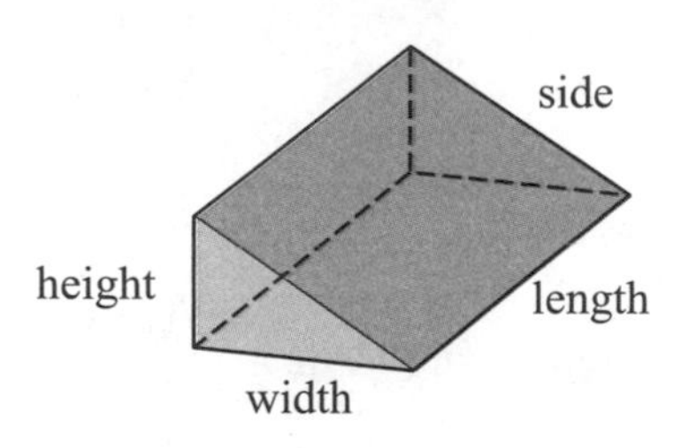

Total Surface Area = $2wl + 2lh + 2wh$

Total Surface Area = $wh + lw + lh + ls$

Examples

1. Find the surface area of the rectangular prism.

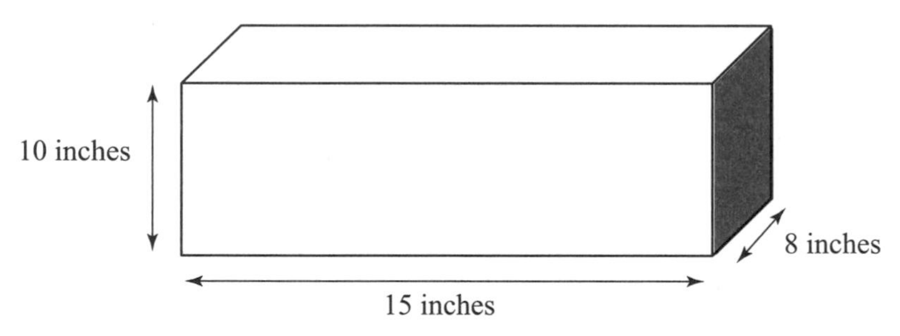

[figure not drawn to scale]

Solution:

Since this is a right rectangular prism, use the formula $2wl + 2lh + 2wh$.

We know that the length = 15 inches, the width = 8 inches, and the height = 10 inches.

Rewrite the formula	$2wl + 2lh + 2wh$
Substitute the values in	$2(8)(15) + 2(15)(10) + 2(8)(10)$
Follow the order of operations	$240 + 300 + 160$
Add	700 square inches

2. Find the surface area of the cylinder. Use 3.14 for π. Round your answer to the nearest tenth.

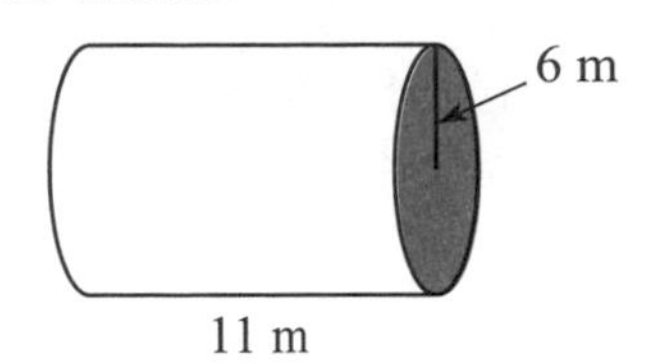

[figure not drawn to scale]

Solution:

Since this is a right circular cylinder, use the formula $2\pi rh + 2\pi r^2$.

We know that the radius = 6 meters, and the height = 11 meters.

Rewrite the formula	$2\pi rh + 2\pi r^2$
Substitute the values in	$2(3.14)(6)(11) + 2(3.14)(6)^2$
Follow the order of operations	$414.48 + 226.08$
Add	640.56 square meters

3. A skateboard ramp is built in the shape of a right triangular prism. Find the surface area of the ramp if the length of the ramp is 15 feet, the width of the ramp is 8 feet, the height of the ramp if 6 feet, and the side of the ramp is 10 feet.

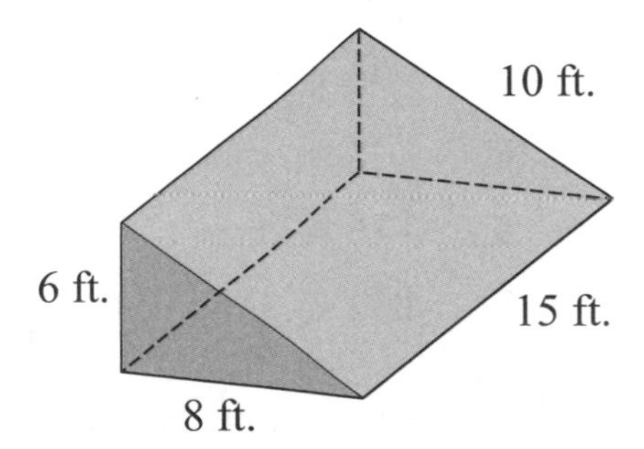

Solution:

Since this is a right triangular prism, use the formula $wh + lw + lh + ls$.

We know $l = 15$, $w = 8$, $h = 6$, and $s = 10$.

Rewrite the original formula	$wh + lw + lh + ls$
Substitute the values in	$8(6) + 15(8) + 15(6) + 15(10)$
Follow the order of operations	$48 + 120 + 90 + 150$
Add	408 square feet

TEST YOUR SKILLS (For answers, see page 264.)

1. Find the surface area of the cylinder below. Use 3.14 for π. Round your answer to the nearest tenth.

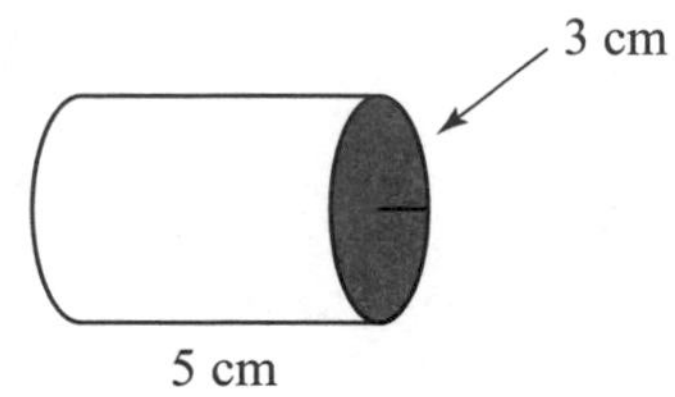

A. 131.9 cm^2
B. 150.7 cm^2
C. 163.4 cm^2
D. 171.3 cm^2

2. Find the surface area of the rectangular prism below.

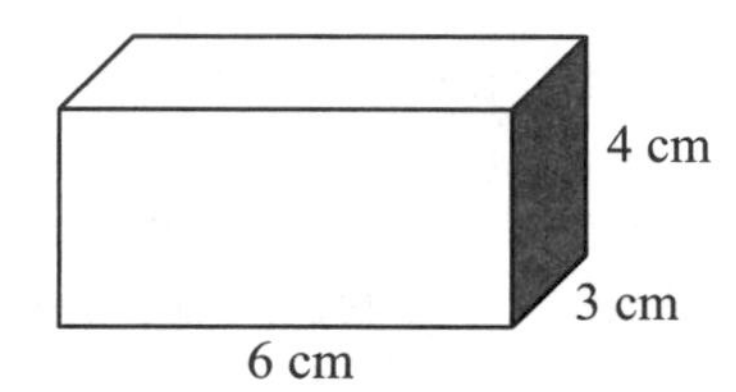

A. 90 sq. cm
B. 108 sq. cm
C. 116 sq. cm
D. 132 sq. cm

3. What is the surface area of a soda can with a radius of 2 inches and a height of 6 inches? Use 3.14 for π. Round your answer to the nearest tenth.

A. 73.4 square inches
B. 87.9 square inches
C. 100.5 square inches
D. 108.2 square inches

4. Kage needs to wrap a birthday present. If she puts the gift in the box below, how much wrapping paper is needed to cover the box?

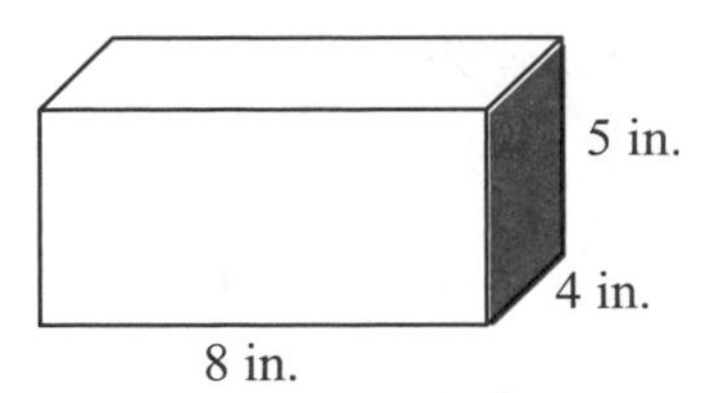

A. 163 in^2

B. 172 in^2

C. 179 in^2

D. 184 in^2

5. Find the surface area of a cylinder if the height is 30 meters and the diameter is 24 meters. Leave your answer in terms of π.

A. 4368π m

B. 4608π m

C. 4712π m

D. 4903π m

6. Danielle decorated her flower vase by wrapping gift wrap around the sides and the bottom of the vase. If the radius of the vase is 4 inches and the height of the vase is 7 inches, find the area of the vase covered by gift wrap. Use 3.14 for π.

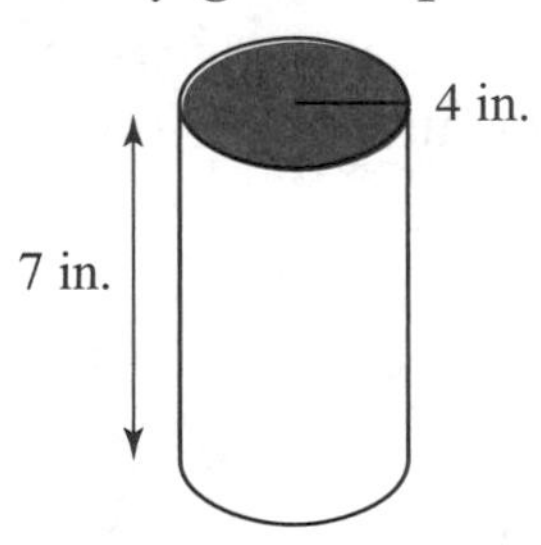

A. 189.45 sq. in.

B. 214.56 sq. in.

C. 226.08 sq. in.

D. 276.32 sq. in.

7. Find the surface area of a cube with a side length of 7 meters.

A. 49 sq. m

B. 84 sq. m

C. 294 sq. m

D. 1764 sq. m

8. Find the surface area of the right triangular prism.

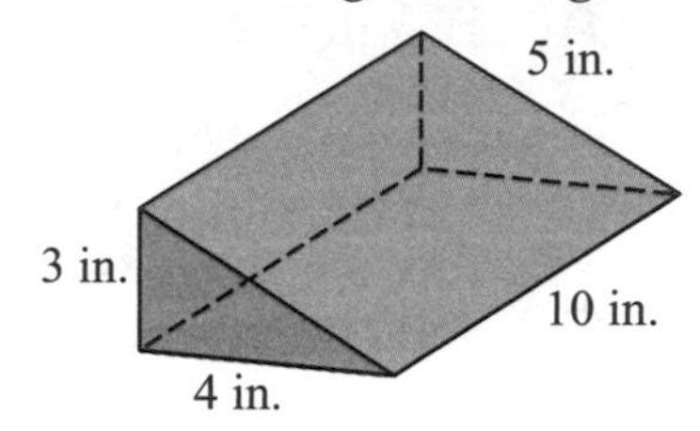

A. 132 square inches

B. 144 square inches

C. 157 square inches

D. 208 square inches

Short Response

9. Design It Boxes has to create a new cereal box for Rice Pops Cereal. The original cereal box was 8 inches long, 6 inches deep, and 10 inches high. The new cereal box has to be 6 inches long, 4 inches deep, and 8 inches high.

How much less material is needed to make the new cereal box compared to the original cereal box? **Show your work.**

Answer __________

Chapter 4

MEASUREMENT

CONVERTING MASS AND WEIGHT

METRIC SYSTEM—UNITS OF MASS

Metric measurement is used by the majority of the world's population. Here is what you need to know:

- The basic unit of mass in the metric system is the **gram** (g).
- A helpful phrase to remember the metric units of mass is "King Henry drank my delicious chocolate milk."

You can use the first letter of each word (khdmdcm) to create the metric staircase. Each letter stands for a metric prefix.

kilo → hecto → deka → meter / gram → deci → centi → milli

Multiply ⟶ (going right, down the staircase)

Divide ⟵ (going left, up the staircase)

Each step is worth 10
Two steps = 100
Three steps = 1000

To change from a larger unit to a smaller unit, multiply.	→	Multiply, go to the right.
To change from a smaller unit to a larger unit, divide.	←	Divide, go to the left.

	EXPLANATION	ANSWER
1. 5 m = _____ cm	2 steps to the right, multiply by 100	500 cm
2. 200 mm = _____ m	3 steps to the left, divide by 1000	0.2 m
3. 5 kg = _____ g	3 steps to the right, multiply by 1000	5000 g
4. 0.06 km = _____ m	3 steps to the right, multiply by 1000	60 m
5. 2.4 cm = _____ mm	1 step to the right, multiply by 10	24 mm
6. 325 mm = _____ m	3 steps to the left, divide by 1000	0.325 m
7. 43 g = _____ kg	3 steps to the left, divide by 1000	0.043 kg
8. .03 g = _____ mg	3 steps to the right, multiply by 1000	30 mg

The metric staircase will help you remember all of the prefixes. For the state test, you are required to know:

1 centimeter = 10 millimeters
1 meter = 100 centimeters = 1000 millimeters
1 kilometer = 1000 meters

1 gram = 1000 milligrams
1 kilogram (kg) = 1000 grams

CUSTOMARY UNITS OF WEIGHT

The basic unit of mass in the customary system is the pound (lb). You can also use a staircase for the customary units. The numbers shown between the staircases are the numbers you needed to convert units.

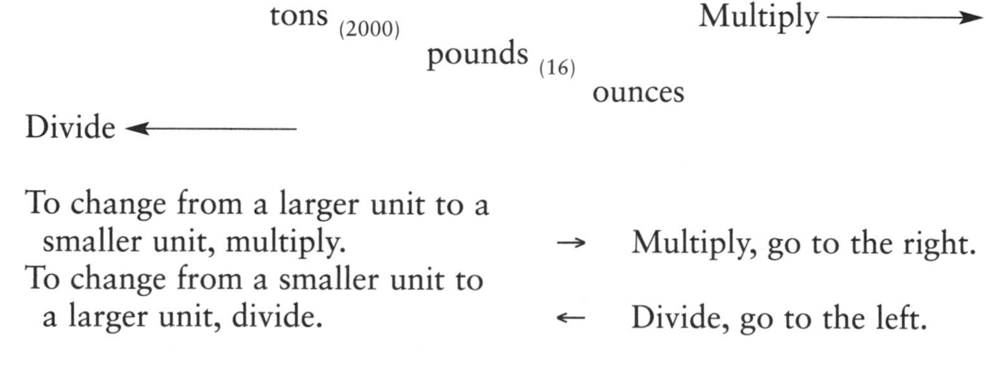

For the state test, you are required to know:

1 pound (lb.) = 16 ounces (oz.)
1 ton (t.) = 2000 pounds

	EXPLANATION	ANSWER
1. 3 t. = _____ lbs.	1 step to the right, multiply by 2000	6000 lbs
2. 48 oz. = _____ lbs.	1 step to the left, divide by 16	3 lbs.
3. 5 lbs. = _____ oz.	1 step to the right, multiply by 16	80 oz.
4. 8000 lbs. = _____ t.	1 step to the left, divide by 2000	4 t.
5. 2 t. = _____ oz.	2 steps to the right, multiply by 2000 and 16	64,000 oz.

TEST YOUR SKILLS (For answers, see page 266.)

1. Which customary unit would you most likely use to measure the mass of a car?

A. grams
B. pounds
C. kilograms
D. tons

2. Which metric unit would you most likely use to measure the length of your bedroom?

A. meters
B. inches
C. grams
D. pounds

3. How do you convert from pounds to ounces?

A. multiply by 16
B. divide by 16
C. multiply by 2000
D. divide by 2000

4. How many centimeters are in 35 millimeters?

A. .035 cm
B. 0.35 cm
C. 3.5 cm
D. 350 cm

5. How many kilometers are in 82 meters?

A. 0.082 km
B. 0.82 km
C. 820 km
D. 82,000 km

6. How many ounces are in 3 pounds?

A. 16 oz.
B. 48 oz.
C. 56 oz.
D. 2000 oz.

7. 3.6 kg = _____ g

A. .0036
B. 36
C. 3600
D. 36,000

8. 10,000 mg = _____ g

A. 0.1
B. 1
C. 10
D. 100

9. 6000 lbs. = _____ tons

A. 2
B. 3
C. 16
D. 2000

10. 60 cm _____ 6 m

A. >
B. <
C. =
D. none of the above

11. 5 g _____ 512 mg

A. >
B. <
C. =
D. none of the above

Short Response

12. You have been studying metric and customary units of measure in math class. Your teacher wrote the following metric measurements on the board:

1200 centimeters	8000 millimeters	35 millimeters
1 kilometers	2 meters	30 meters

Part A.

Order the measurements from least to greatest. Show your work.

Answer

_____ _____ _____ _____ _____ _____

Part B.

Your teacher only gave you six measurements to put in order. If she decided to give you a seventh measurement to add to the list, would 8 pounds be appropriate to add in Part A? Explain your answer.

____

____

____

CONVERTING CAPACITY AND VOLUME

METRIC SYSTEM—UNITS OF CAPACITY

The basic unit of capacity in the metric system is liters (L).

You can once again use the phrase "King Henry drank my delicious chocolate milk" to help you with units of capacity. The staircase will work for meters, liters, and grams.

Just remember: meters = metric unit of length
grams = metric unit of weight
liters = metric unit of capacity (liquid)

In this section, we will review capacity.

kilo — Multiply ⟶
hecto
deka
meter
gram deci
liter centi
Divide ⟵ milli

Each step is worth 10
Two steps = 100
Three steps = 1000

To change from a larger unit to a smaller unit, multiply.	→	Multiply, go to the right.
To change from a smaller unit to a larger unit, divide.	←	Divide, go to the left.

For the state test, you will be required to know these conversions:

> 1 liter (L) = 1000 milliliters (mL)
> 1 kiloliter (kL) = 1000 liters

	EXPLANATION	ANSWER
1. 2 L = _____ mL	3 steps to the right, multiply by 1000	2000 mL
2. 5 kL = _____ L	3 steps to the right, multiply by 1000	5000 L
3. 6200 L = _____ kL	3 steps to the left, divide by 1000	6.2 kL
4. 63 mL = _____ L	3 steps to the left, divide by 1000	0.063 L

CUSTOMARY UNITS OF VOLUME

The basic unit of volume in the customary system is quart (qt).

Here is a staircase to help you remember how to convert. The numbers shown between the staircases are the numbers you need to convert units.

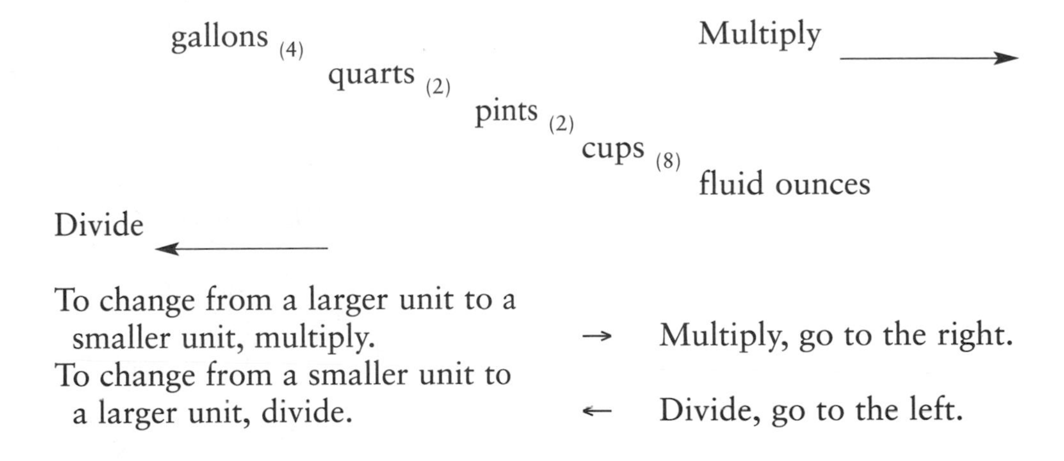

To change from a larger unit to a smaller unit, multiply.	→	Multiply, go to the right.
To change from a smaller unit to a larger unit, divide.	←	Divide, go to the left.

For the state test, you will be required to know these conversions:

1 cup (c.) = 8 fluid ounces (fl. oz.)
1 pint (pt.) = 2 cups
1 quart (qt.) = 2 pints
1 gallon (gal.) = 4 quarts

		EXPLANATION	ANSWER
1.	2 gal. = ______qts.	1 step to the right, multiply by 4	8 quarts
2.	88 fl. oz. = _____ cups	1 step to the left, divide by 8	11 cups
3.	36 pts. = _____ qts.	1 step to the left, divide by 2	18 quarts
4.	3 pts. = _____ c.	1 step to the right, multiply by 2	6 cups
5.	3 gal. = _____ pts.	2 steps to the right, multiply by 4 and 2	24 pints
6.	96 fl. oz. = _____ pts.	2 steps to the left, divide by 8 and 2	6 pints

TEST YOUR SKILLS (For answers, see page 268.)

1. 425 mL = _____ L

   A. 0.425
   B. 4.25
   C. 42.5
   D. 4250

2. You want to make pancakes for breakfast. The recipe says that you will need 2 cups of milk. How many fluid ounces will this be?

   A. 2 fl. oz.
   B. 8 fl. oz.
   C. 12 fl. oz.
   D. 16 fl. oz.

3. The gas tank in a fishing boat will hold 16 quarts of gasoline. What is this measurement in gallons?

   A. 2 gal.
   B. 4 gal.
   C. 8 gal.
   D. 32 gal.

4. Oscar bought 2 quarts of ice cream. How many pints of ice cream is this?

   A. 2 pts.
   B. 4 pts.
   C. 8 pts.
   D. 16 pts.

5. You have 4 liters of soda. How much soda do you have in milliliters?

   A. 0.4 mL
   B. 40 mL
   C. 400 mL
   D. 4000 mL

6. Rainer bought 3 pints of ice cream to make an ice cream cake. The recipe for the ice cream cake measures the ice cream in cups. If Rainer has to use all 3 pints of ice cream for the recipe, how many cups of ice cream is this?

   A. 2 c.
   B. 5 c.
   C. 6 c.
   D. 24 c.

7. How many cups are in 4 quarts?

A. 2 c.

B. 4 c.

C. 16 c.

D. 20 c.

8. Kevin put 3 gallons of gasoline in his 4-wheeler. How many quarts of gasoline is this?

A. 4 qts.

B. 7 qts.

C. 12 qts.

D. 16 qts.

9. How many liters are in 0.5 kiloliters?

A. .005 L

B. 5 L

C. 50 L

D. 500 L

10. What is the basic unit of measure in the metric system for mass?

A. pound

B. gram

C. liter

D. meter

Short Response

11. Convert 12.5 liters to kiloliters.

Part A.

Show your work.

Answer __________

Part B.

On the lines below, explain why you should divide when changing from liters to kiloliters.

____

____

CHOOSING THE APPROPRIATE TOOL AND JUSTIFYING THE REASONABLENESS OF THE MASS OF AN OBJECT

There are several different types of scales used for measuring the mass of an object. A scale is used to tell you how much something weighs.

Examples of different scales are:

BALANCE

A balance is used to compare the mass of two objects.

Examples:

A pile of nickels compared to a pile of quarters
A pile of paper clips compared to a pile of thumbtacks
A CD compared to a calculator

PRODUCE SCALE

A produce scale is used to measure the mass of food (produce).

Examples:

A bag of grapes
A bag of apples
A bunch of bananas

BATHROOM SCALE

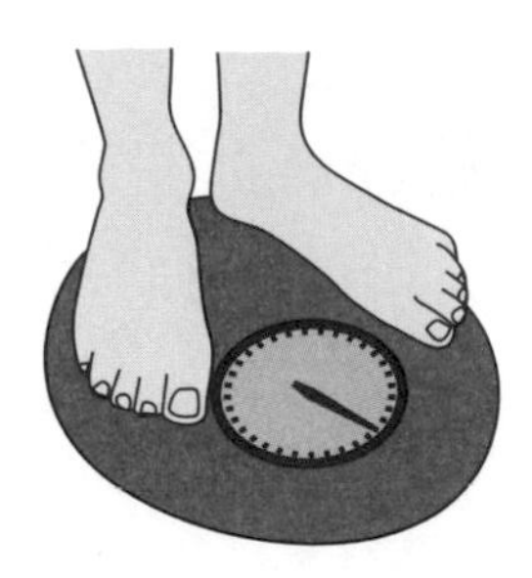

A bathroom scale is used to measure the mass of a person.

Examples:

You!
A baby

TRUCK SCALE

A truck scale is used to measure the mass of heavy objects.

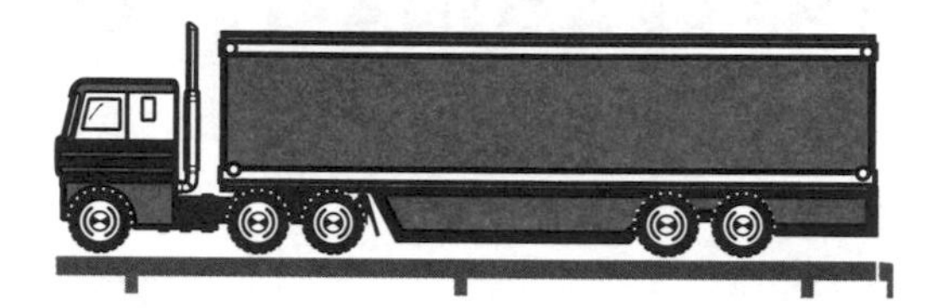

Examples:

A car
A truck
An elephant

TEST YOUR SKILLS (For answers, see page 269.)

1. Which scale should be used to compare the mass of a cell phone and an MP3 player?

   A. balance

   B. produce scale

   C. bathroom scale

   D. truck scale

2. Which scale should be used to measure the mass of a tractor?

   A. balance

   B. produce scale

   C. bathroom scale

   D. truck scale

3. Which scale is used when you are comparing the mass of two objects?

   A. balance

   B. produce scale

   C. bathroom scale

   D. truck scale

4. In science class, you are finding the mass of different objects. You know that a box of pencils has a mass of 10 grams. If 8 boxes of pencils equals the mass of your calculator, how much does your calculator weigh?

   A. 8 g

   B. 80 g

   C. 100 g

   D. 800 g

5. You want to know the weight of your puppy, so you and your puppy stand on the bathroom scale. The scale says you and the puppy weigh 110 pounds. What is a reasonable estimate of how much you weigh if you are not holding the puppy?

   A. 12 lbs.

   B. 30 lbs.

   C. 90 lbs.

   D. 130 lbs.

Short Response

6. One pack of gum weighs 5 grams. There are five pieces of gum in a pack. Your MP3 player weighs 35 grams.

   How many pieces of gum would equal the mass of your MP3 player? Show your work.

   Answer __________

ESTIMATING SURFACE AREA

The surface area of a three-dimensional shape is the area of the outside surface of the shape. To find the total surface area, you need to find the area of each side of the shape and add them all together.

To estimate surface area means that you are finding an *approximate* calculation of the surface, *not* the exact surface area.

The important concept to remember if you are estimating surface area is *round the decimals first, then find the surface area.*

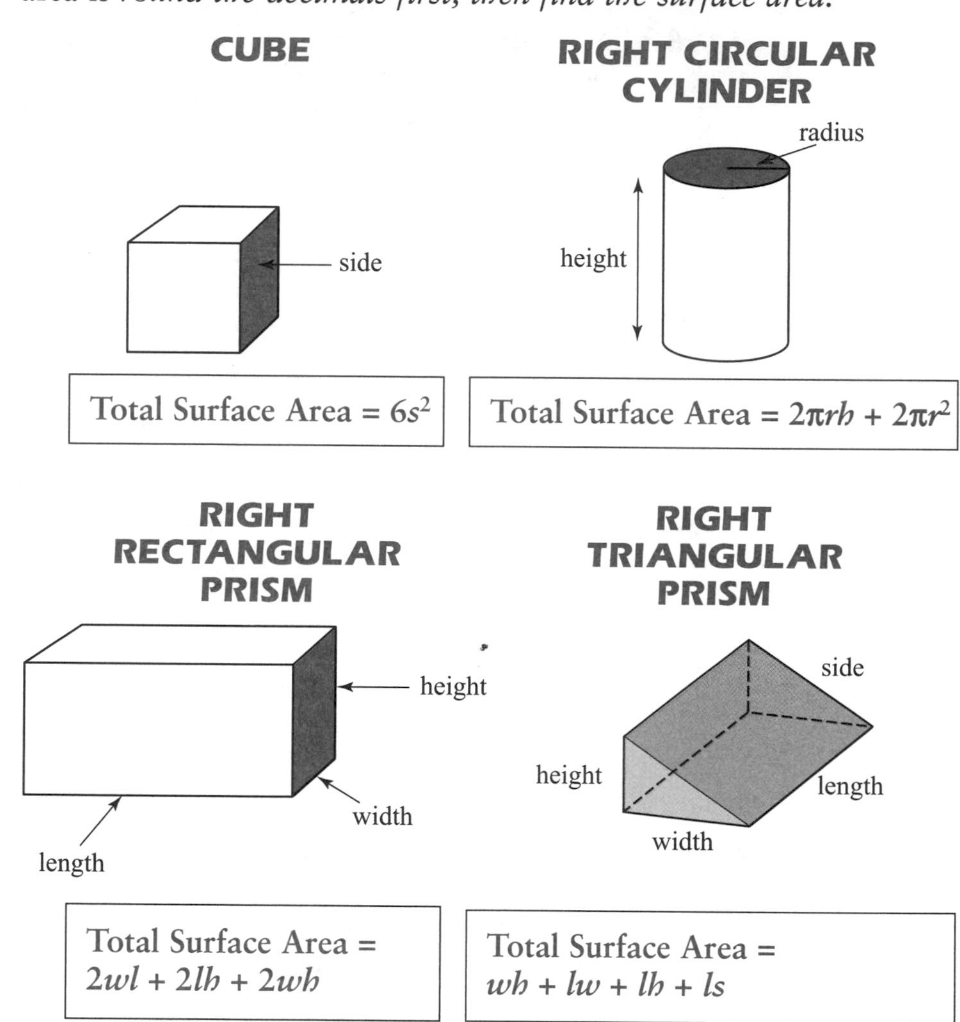

Examples

1. Use estimation to calculate the total surface area of the shoebox.

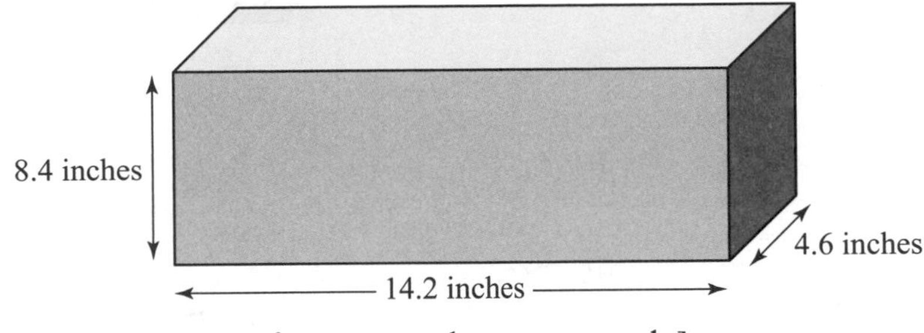

[figure not drawn to scale]

Solution:

Because the problem tells you to use estimation, first round each side to the nearest whole number.

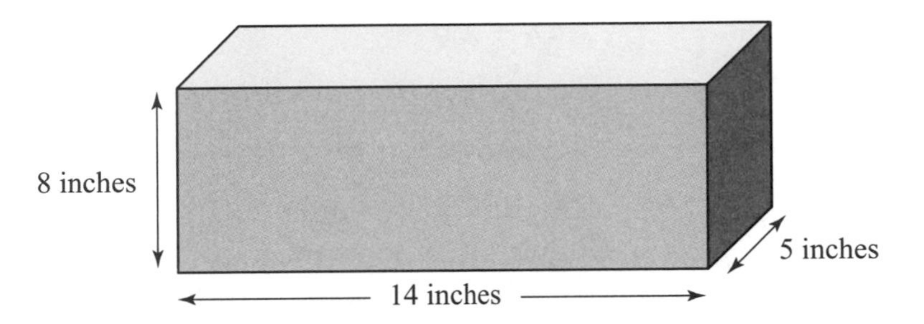

Since this is a right rectangular prism, use the formula $2wl + 2lh + 2wh$.

We know that the length = 14 inches, the width = 5 inches, and the height = 8 inches.

Rewrite the formula	$2wl + 2lh + 2wh$
Substitute your values in	$2(5)(14) + 2(14)(8) + 2(5)(8)$
Follow the order of operations	$140 + 224 + 80$
	444 square inches

2. What is an estimate for the surface area of the following cylinder?

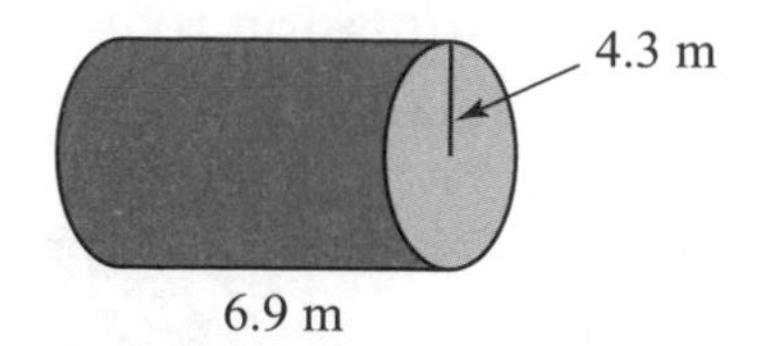

[figure not drawn to scale]

Solution:

First round the radius and the height to the nearest whole number.

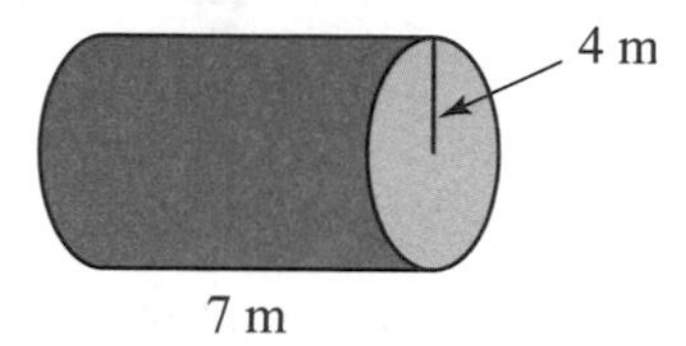

Since this is a right circular cylinder, use the formula $2\pi rh + 2\pi r^2$.

Remember to use 3 for π since we are estimating.

We know that the radius = 4 meters and the height = 7 meters.

Rewrite the formula	$2\pi rh + 2\pi r^2$
Substitute your values in	$2(3)(4)(7) + 2(3)(4)^2$
Use the order of operations	$168 + 96$
	264 square meters

TEST YOUR SKILLS (For answers, see page 269.)

1. Use estimation to find the surface area of the cube.

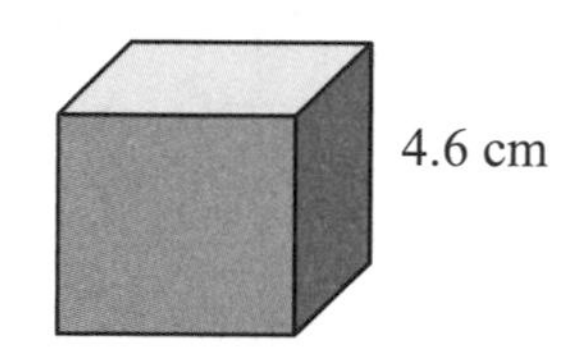

[figure not drawn to scale]

A. 60 cm^2

B. 80 cm^2

C. 150 cm^2

D. 180 cm^2

2. What is an estimate of the surface area of the given rectangular prism?

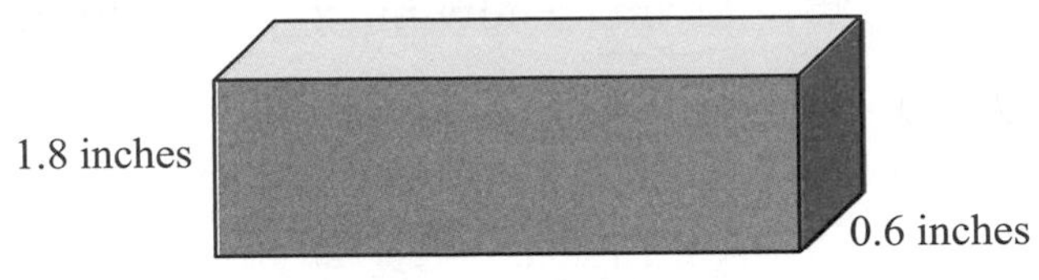

[figure not drawn to scale]

A. 16 in^2

B. 72 in^2

C. 114 in^2

D. 128 in^2

3. Estimate the surface area of a right circular cylinder with a radius of 3.6 centimeters and a height of 8.1 centimeters.

A. 186 cm^2

B. 227 cm^2

C. 277 cm^2

D. 288 cm^2

4. What is an estimate for the total surface area of a cube with a side length of 3.8 inches?

A. 36 sq. in.

B. 48 sq. in.

C. 82 sq. in.

D. 96 sq. in.

5. A right triangular prism has a width of 2.1 centimeters, a length of 6.3 centimeters, a height of 4.4 centimeters, and a side of 5.1 centimeters. Which is the best estimate of the surface area of the triangular prism?

A. 78 sq. cm

B. 82 sq. cm

C. 91 sq. cm

D. 102 sq. cm

6. Owen is painting a box that is the shape of a cube for an art project. The length of the side of the cube is 12.3 inches. Estimate how much paint Owen will need to cover the entire box.

A. 144 in^2

B. 864 in^2

C. 900 in^2

D. 5184 in^2

7. Tracy bought a big tin of assorted popcorn. The popcorn tin is a cylinder container. The container has a diameter of 6 inches and a height of 10.2 inches. Which is the best estimate of the surface area?

A. 204 sq. in.

B. 234 sq. in.

C. 396 sq. in.

D. 414 sq. in

Short Response

8. A rectangular vase is 11.2 inches tall, 4.7 inches wide, and 2.7 inches deep. Find the surface area of the vase.

Part A.

Show your work.

Answer __________

Part B.

On the lines below, explain why the formula $wl + 2lh + 2wh$ is used instead of $2wl + 2lh + 2wh$ to find the surface area of the vase.

____

____

____

CONSTRUCTING CENTRAL ANGLES

A central angle is an angle that intersects a circle in two places and its vertex is the center of the circle.

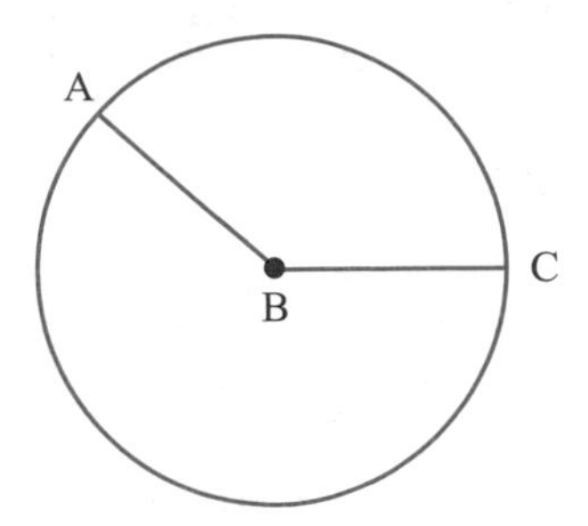

Examples

1. Name the central angle.

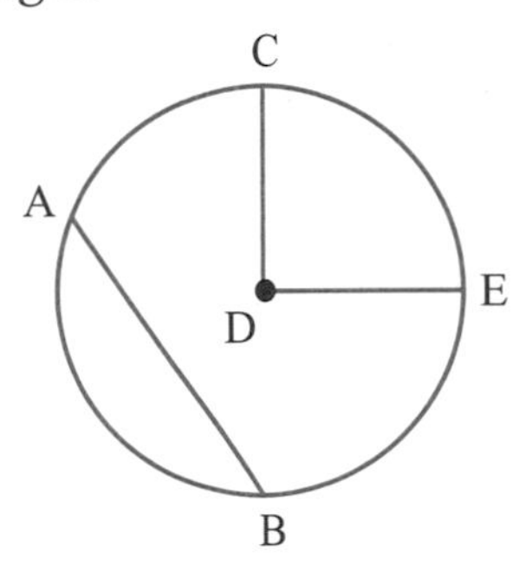

Solution: The name of the angle is $\angle CDE$ or $\angle EDC$. The vertex of the angle has to be in the middle when naming the angle.

2. What is the measure of the central angle?

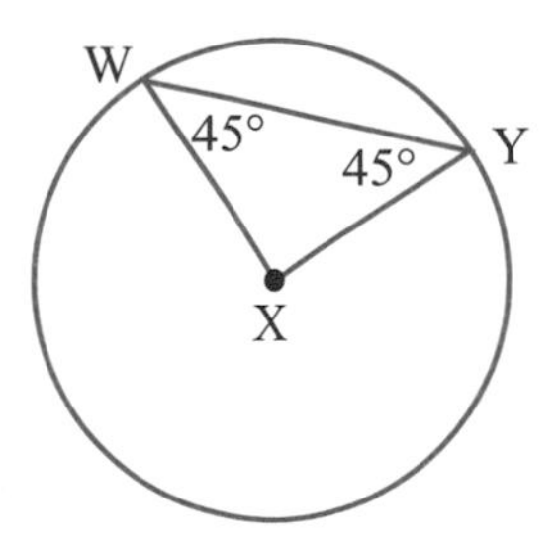

Solution: $\angle WXY$ is the central angle. There are 180° in a triangle. Since the two angles given add up to 90°, $180 - 90 = 90$. $\angle WXY = 90°$.

CONSTRUCTING CENTRAL ANGLES WITH A PROTRACTOR

A compass is an instrument used to draw circles. A protractor is a tool you can use to draw and measure angles. In this section, you will need to be able to draw an angle in a circle when the circle has been provided for you.

Example

1. Construct a 120° angle and label the angle XYZ.

   Solution:

   Position your protractor correctly on the center of the circle. Using the bottom of the protractor, draw a straight line from the center of the circle to a point on the circle. This is your radius. With your protractor correctly on the radius, start from 0° on the protractor and measure to 120°. Mark where 120° would be on the circle.
   Using your protractor as a straightedge, draw the new radius. Label the angle XYZ.

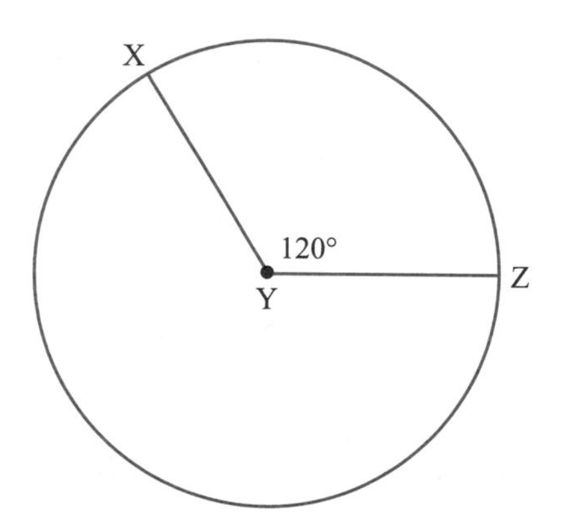

TEST YOUR SKILLS (For answers, see page 270.)

1. Name the central angle.

   A. $\angle AEC$
   B. $\angle BEC$
   C. $\angle CDE$
   D. $\angle AEB$

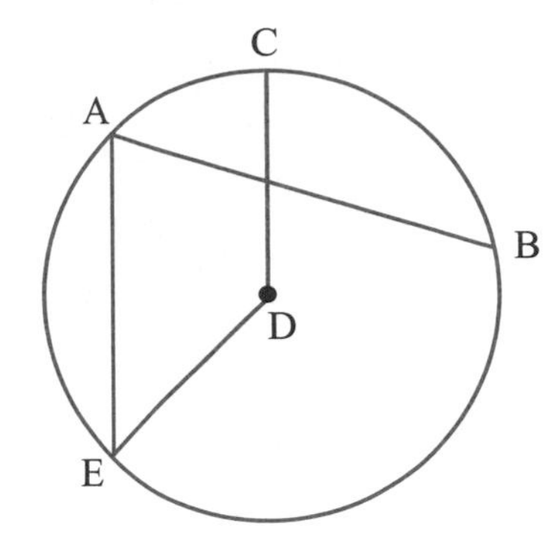

2. Name the central angle.

   A. $\angle CEF$
   B. $\angle GHJ$
   C. $\angle JGH$
   D. $\angle CEG$

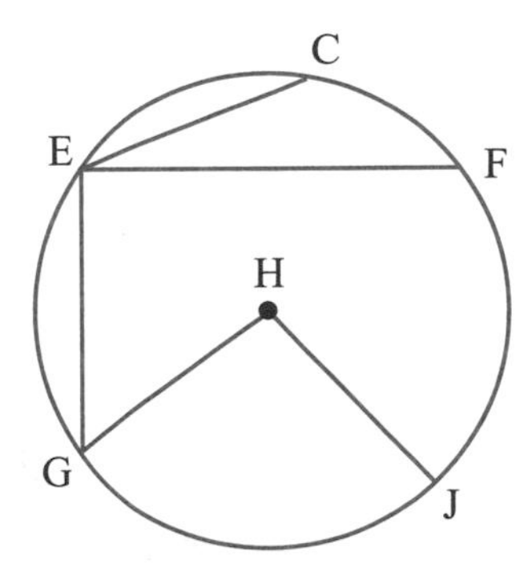

3. Which angle is a central angle with a measure of 30°?

A.

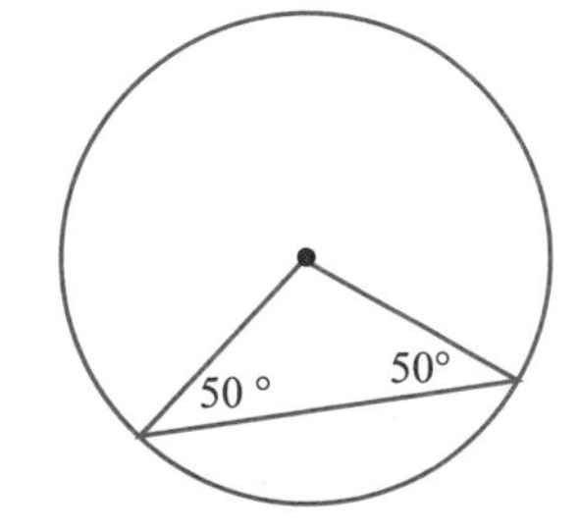

B.

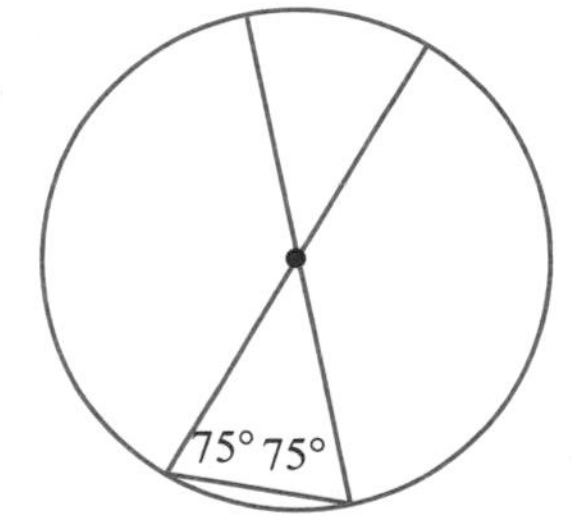

C.

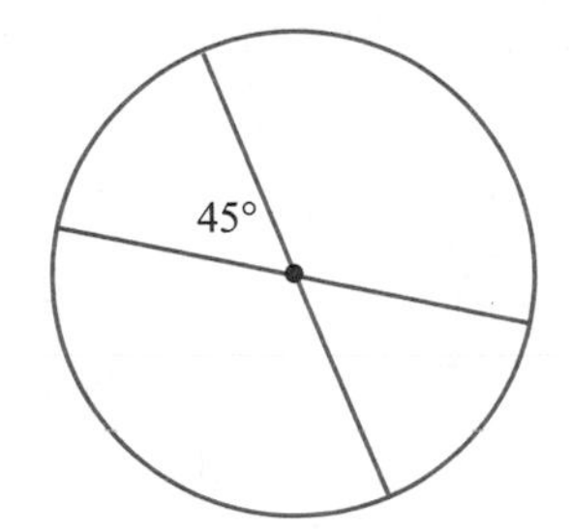

D.

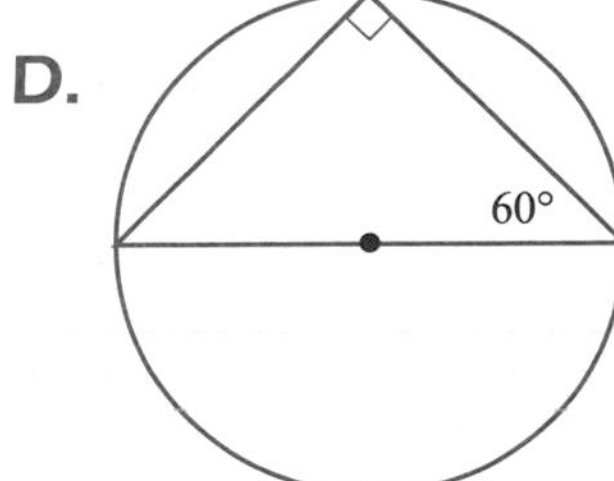

4. Which angle is a central angle with a measure of 60°?

A.

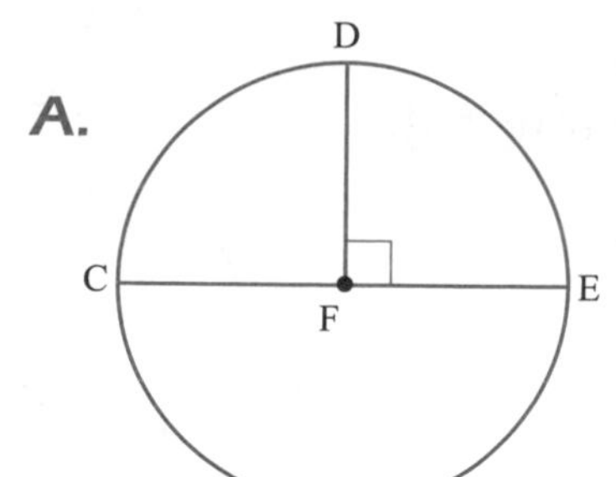

B.

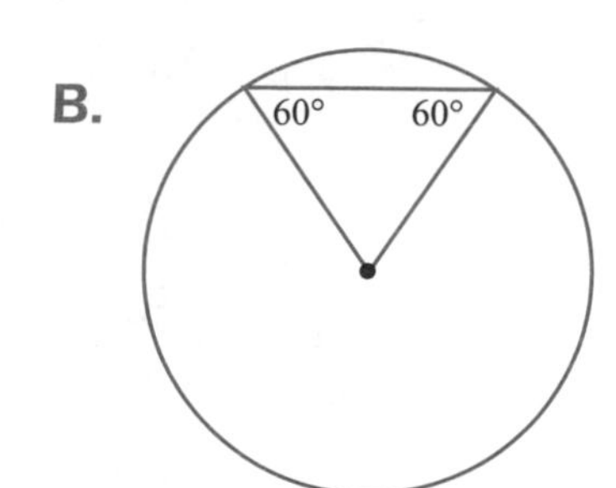

C.

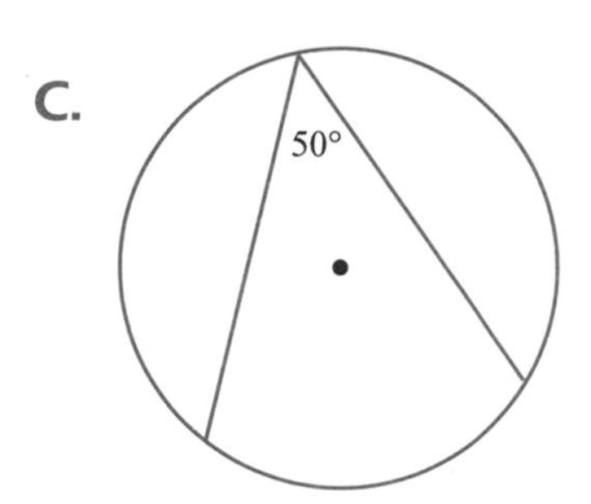

D.

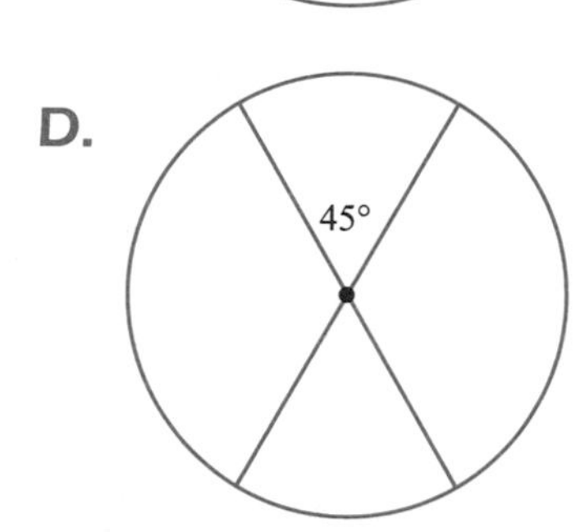

5. Which circle does *not* contain a central angle?

A.

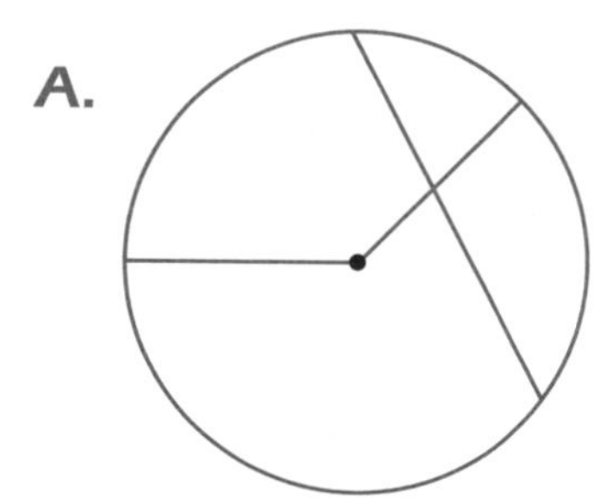

B.

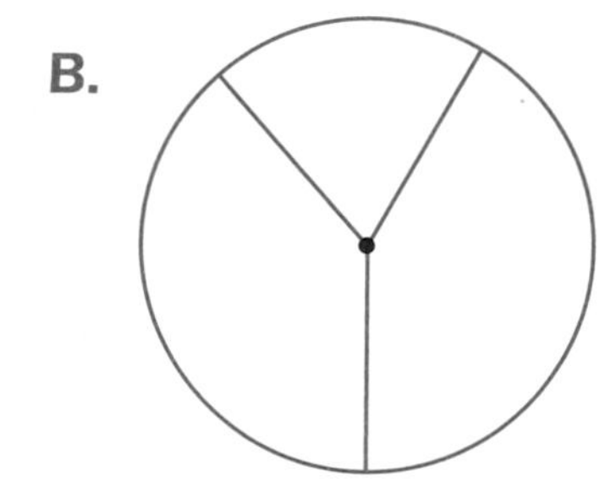

C.

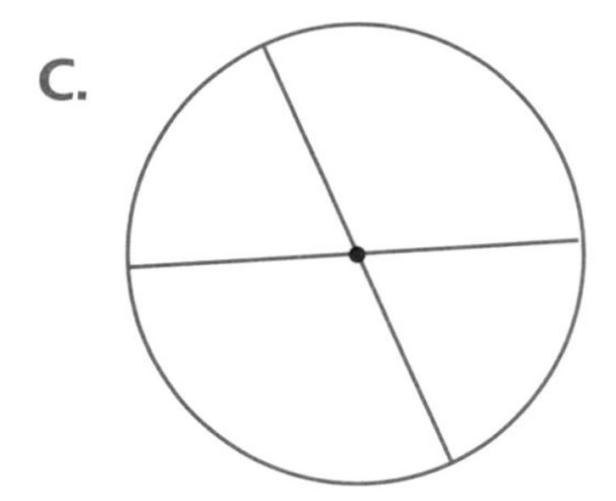

D.

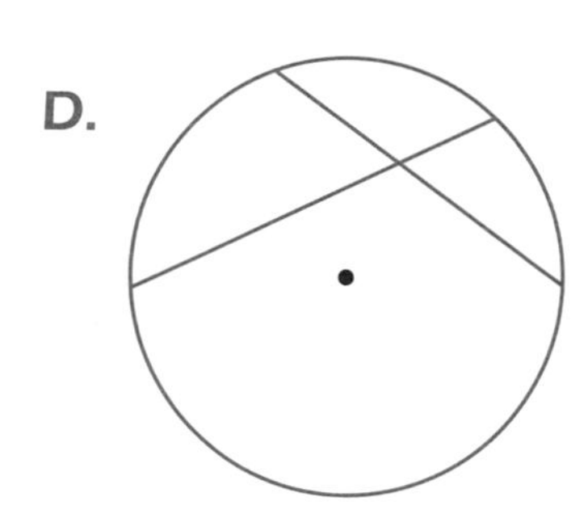

Short Response

6. Draw a central angle with a measure of 45°. Label the angle *ABC*.

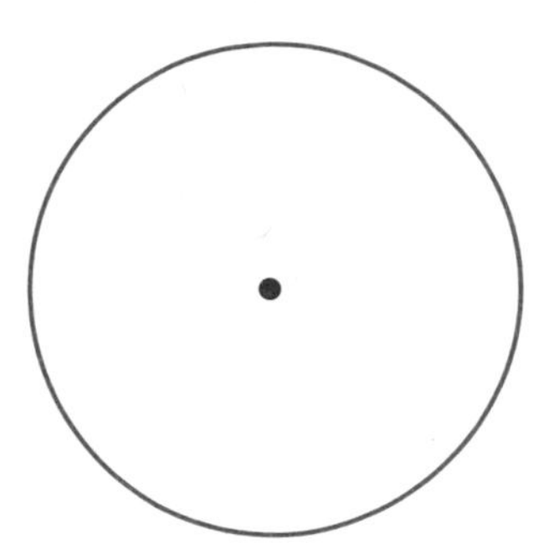

RELATIVE ERROR

Have you ever used a protractor in math class to measure angles? You measure an angle of 38° and your friend measures the same angle and gets 37°?

It is often hard to get an exact measurement. Therefore, to give an exact measurement, the measurement is given as a range. The range is the actual measurement plus or minus the **error** in measurement.

Relative error is the ratio of error in measurement to the approximate or "true" measurement.

$$\frac{\text{Error in measurement}}{\text{Approximate measurement}}$$

Example

In science class, Jade determined that the mass of a rock was 7.2 grams. Jade's teacher is allowing measurements to within ± 0.2 grams to be correct.

What is the range of the mass of Jade's rock?

Solution:

To find the range, write the actual length ± the error.
7.2 – 0.2 = 7.0
7.2 + 0.2 = 7.4
The range is 7.0–7.4 grams.

Example

What is the relative error of the mass of a paper clip if the actual weight is 2 grams $\pm \frac{1}{2}$ gram?

Solution:

The relative error is

$$\frac{\text{Error in measurement}}{\text{Approximate measurement}}$$

$$\frac{\left(\frac{1}{2}\right)}{2} = \frac{1}{2 \div 2}$$

$$= \frac{1}{2} \cdot \frac{1}{2}$$

$$= \frac{1}{4}$$

The relative error is $\frac{1}{4}$.

TEST YOUR SKILLS (For answers, see page 271.)

1. What is the relative error of the length of a shoelace that is 10 inches long ± 0.4 inches?

   A. $\frac{1}{25}$

   B. $\frac{1}{4}$

   C. $\frac{25}{1}$

   D. $\frac{1}{10}$

2. What is the range for the weight of a newborn baby that is 7.2 pounds with an error of measurement of 0.4 pound?

   A. 6.8 lbs. to 7.6 lbs.

   B. 6.8 lbs. to 11.4 lbs.

   C. 7.6 lbs. to 11.4 lbs.

   D. 9.4 lbs. to 11.4 lbs.

3. Tonya needs 8.5 yards of fabric to make costumes for the school musical. In case she makes a mistake, she is going to order an extra 2 yards of fabric. Which is the relative error of the length of the fabric?

A. $\frac{17}{4}$

B. $\frac{1}{25}$

C. $\frac{1}{5}$

D. $\frac{4}{17}$

4. What is the range in measurement for ribbon that is 5.4 inches long with an error of measurement of $\frac{1}{4}$ inch?

A. 4.9 in. to 5.1 in.

B. 4.9 in. to 5.65 in.

C. 5.15 in. to 5.65 in.

D. 5.15 in. to 5.9 in.

5. Which could be the weight of a textbook that has a range of 4.6 pounds to 5.4 pounds?

A. 4 lbs. ± 0.6 lbs.

B. 4.6 lbs. ± 0.8 lbs.

C. 5 lbs. ± 0.4 lbs.

D. 5 lbs. ± 0.6 lbs.

6. Which is the relative error of a basketball hoop that is 10 feet ± 0.2 feet?

A. $\frac{1}{25}$

B. $\frac{1}{50}$

C. $\frac{1}{10}$

D. $\frac{1}{5}$

Short Response

7. Jude ordered 1.5 pounds of ham from the deli. The butcher said he went over $\frac{1}{4}$ of a pound.

Part A.

What is the range in measurement for the weight of the ham? Show your work.

Answer __________

Part B.

What is the relative error for the amount of ham Jude ordered? Show your work.

Answer __________

Chapter 5

PROBABILITY

LISTING OUTCOMES OF COMPOUND EVENTS

What do you think of when you hear the word probability? Maybe tossing a die, flipping a coin, picking an ace from a deck of cards, or spinning a spinner. We can perform simple experiments using one or more of these items.

EXAMPLES

1. If you flip a coin, how many possible outcomes are there? What are they?

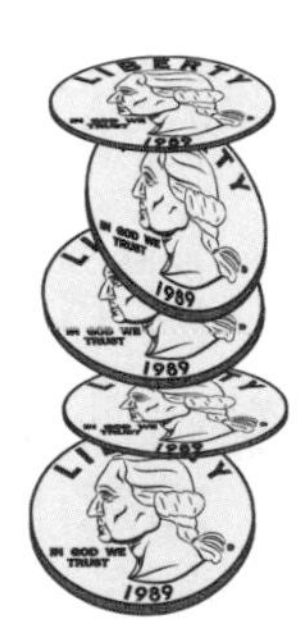

Solution: There are two possible outcomes: heads or tails.

2. If you roll a die, how many possible outcomes are there? What are they?

Solution: There are six possible outcomes: 1, 2, 3, 4, 5, or 6.

3. How many possible outcomes would there be if you rolled the die and flipped the coin?

Solution:

You can use a **tree diagram** to show all the possible outcomes for the event.

Step 1: List the outcomes for the die: 1, 2, 3, 4, 5, 6. Make this the first part of the tree diagram.

Step 2: List the outcomes for the coin: H, T. Make this the second branch of the tree diagram.

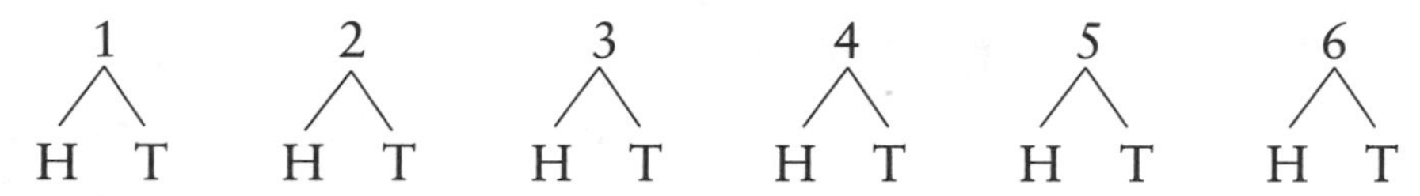

Step 3: Trace each branch to list all the possible outcomes.

The possible outcomes are: 1H, 1T, 2H, 2T, 3H, 3T, 4H, 4T, 5H, 5T, 6H, 6T

There are 12 possible outcomes if you roll a die and flip a coin.

4. List the possible outcomes for flipping a coin three times.

Step 1: List the outcomes for tossing the coin the first time: H, T. Make this the first branch.

Step 2: List the outcomes for tossing the coin a second time: H, T. Make this the second branch.

Step 3: List the outcomes for tossing the coin a third time: H, T. Make this the third branch.

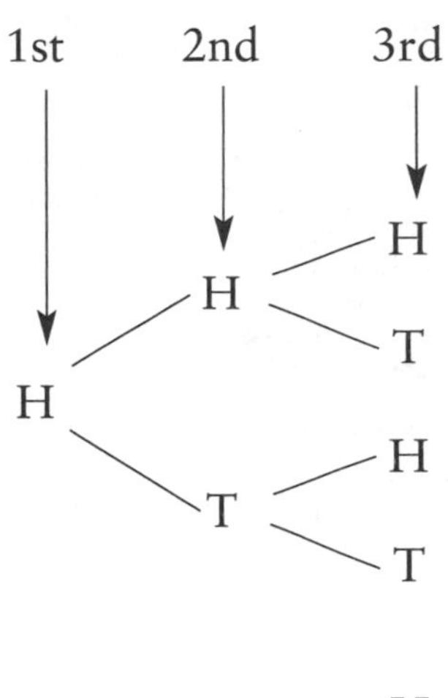

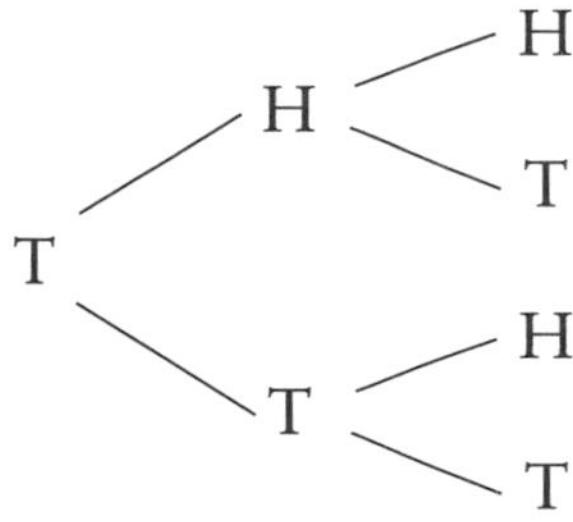

Step 4: Trace each branch to list all possible outcomes:
HHH, HHT,HTH, HTT, THH, THT, TTH, TTT
There are eight possible outcomes if you flip a coin three times.

Note

A tree diagram can be drawn horizontally (Example 3) or vertically (Example 4). It is your choice!

TEST YOUR SKILLS (For answers, see page 272.)

1. How many possible outcomes are there for spinning a spinner with equal sectors labeled A through D and tossing a quarter?

   A. 5
   B. 6
   C. 7
   D. 8

2. You want to buy a new cell phone. You have the choice between a flip phone or a camera phone. They are available in red, blue, chocolate, or yellow and you have to choose between 500 minutes or 1000 minutes. How many different cell phone choices do you have?

A. 8

B. 10

C. 16

D. 20

3. At Cookie's Sub Shop, you can order a small, medium, or large sub. Your meat choices are ham or turkey. Your cheese choices are American or Swiss. If you choose one size, one meat, and one cheese, how many possible subs are there?

A. 3

B. 12

C. 21

D. 54

4. List the possible outcomes if you have the choice between a bagel, muffin, or doughnut for breakfast and milk, orange juice, or coffee to drink.

____

5. List the possible outcomes for spinning a spinner with three equal sectors labeled A, B, and C and rolling a cube numbered 1 through 6.

____

6. List the possible outcomes for rolling two dice each numbered 1 through 6.

____

7. An experiment consists of spinning a spinner with three equal sectors labeled X, Y, and Z and spinning a second spinner with four equal sectors labeled 1, 2, 3, and 4. List all of the possible outcomes.

____

8. Your soccer coach wants to order new jerseys. List the possible outcomes if the jerseys are available in red, white, blue, or grey and they come in sizes small, medium, or large.

____

Short Response

9. Brent designed an experiment consisting of rolling a cube numbered 1 through 6 and choosing one of the following cards.

A B C

Part A.

List the possible outcomes for Brent's experiment. Show your work.

Answer __________

Part B.

How many total outcomes are there?

Answer __________

THE FUNDAMENTAL COUNTING PRINCIPLE

For breakfast, you have the choice between eggs, waffles, or pancakes with sausage, bacon, or ham. How many different breakfast choices do you have?

One method to solve this problem is to use a **tree diagram.**

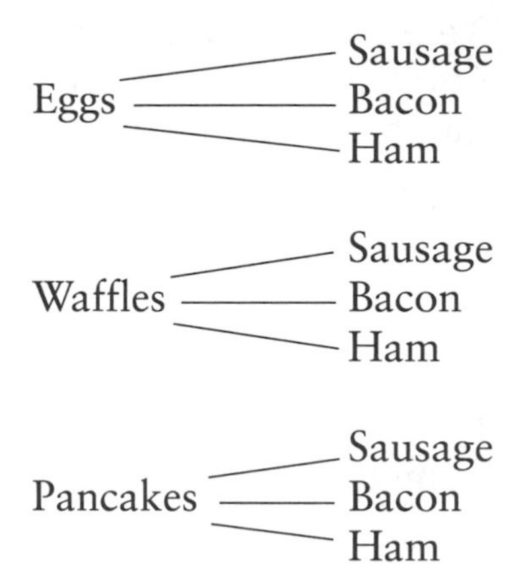

There are nine possible choices for breakfast: ES, EB, EH, WS, WB, WH, PS, PB, PH.

A second method is to use the **fundamental counting principle.** The fundamental counting principle states that if one event happens in m ways, and another event happens in n ways, then the two events happen in $m \times n$ ways. *In other words, multiply the number of outcomes for each event together.*

For our breakfast example, there are three choices for the main dish and three choices for the side dish.

Apply the counting principle: $3 \times 3 = 9$. There are nine possible choices for breakfast.

Examples

1. Three dice are rolled. How many possible outcomes are there?

Solution: There are 6 outcomes for the first die: 1, 2, 3, 4, 5, 6
There are 6 outcomes for the second die: 1, 2, 3, 4, 5, 6
And there are 6 outcomes for the third die: 1, 2, 3, 4, 5, 6

Apply the counting principle: $6(6)(6) = 216$

There are 216 possible outcomes when you roll three dice.

Strategy: Draw a picture of three dice ☐ ☐ ☐

Write the outcomes for each die 6 × 6 × 6

Multiply 216 outcomes

2. An experiment consists of spinning a spinner with four equal sectors labeled A, B, C, and D and flipping a coin. How many outcomes are possible?

Solution: There are 4 outcomes for the spinner: A, B, C, D
There are 2 outcomes for the coin: H or T

Apply the counting principle: 4(2) = 8

There are 8 possible outcomes.

Strategy: Draw a picture of the spinner and the coin ◯ ◯

Write out the outcomes for each object 4 × 2

Multiply 8 outcomes

3. A security code for your cell phone uses four digits. The possible digits for the code are 0–9. How many possible combinations are there for a security code if the digits can be repeated?

Solution: Make a picture. The security code can only be four digits long.

____ ____ ____ ____

For each "spot" you can use the digits 0–9. That means there are 10 choices for each digit of the security code because digits can be repeated.

$\underline{10} \cdot \underline{10} \cdot \underline{10} \cdot \underline{10} = 10{,}000$

There are 10,000 possible combinations for the security code.

Be happy you know the fundamental counting principle—imagine making this tree diagram!

TEST YOUR SKILLS (For answers, see page 275.)

1. An experiment consists of throwing a die twice. How many possible outcomes are there?

   A. 6
   B. 12
   C. 18
   D. 36

2. You want to create a security code for your bank account. The code has to be four digits long. The possible digits are 0–9 and the numbers can be repeated. How many different security codes are possible?

   A. 1000
   B. 5040
   C. 6561
   D. 10,000

3. Brittany has five coins. How many outcomes are possible if Brittany tosses all five coins?

   A. 10
   B. 25
   C. 32
   D. 50

4. The chart below shows the dinner choices at a local restaurant.

Appetizers	Main Dish	Dessert
Chicken wings	Steak	Hot fudge sundae
Mozzarella sticks	Chicken sandwich	Apple pie
Nachos	Spaghetti	
	Fish fry	

How many possible outcomes are there if you choose one appetizer, one main dish, and one dessert?

A. 3

B. 9

C. 12

D. 24

5. An experiment consists of tossing a penny, a dime, and a nickel and rolling a die numbered 1 through 6. How many outcomes are possible?

A. 12

B. 18

C. 36

D. 48

6. The registration sticker on a boat is made up of two letters followed by four digits. How many possible combinations are there for a registration sticker if the letters can only be consonants and the digits can be 0–9? The consonants and digits can be repeated.

A. 82

B. 676,000

C. 4,410,000

D. 4,435,236

7. How many four letter words can be made from the letters P-Y-R-A-M-I-D if the word must begin and end with a vowel and all of the letters can be repeated?

A. 14

B. 100

C. 196

D. 225

8. An experiment consists of spinning a spinner with equal sectors labeled red, white, blue, and green and rolling a cube numbered 1 through 6. How many outcomes are possible?

A. 10

B. 24

C. 48

D. 90

Short Response

9. A cube numbered 1 through 6 is rolled four times. Using the fundamental counting principle, how many possible outcomes are there? **Show your work.**

Answer ________

10. At Vanilla Swirl, you have the choice between chocolate, vanilla, or bubblegum ice cream. They have three different types of cones: sugar, waffle, or regular. Your sprinkle choices are chocolate or rainbow.

Part A.

Using the fundamental counting principle, how many possible outcomes are there if you have to choose one flavor of ice cream, one type of cone, and one topping? **Show your work.**

Answer ________

Part B.

Explain why your answer is correct.

____

____

____

PROBABILITY OF DEPENDENT EVENTS

Probability—the chance that an event will occur. It is the ratio of the number of favorable outcomes to the number of possible outcomes.

$$\text{Probability} = \frac{\text{Number of favorable outcomes}}{\text{Number of possible outcomes}}$$

In probability, there are independent and dependent events.

Example

Find the probability of rolling a 4 on a cube numbered 1 through 6, and spinning an even number on the spinner.

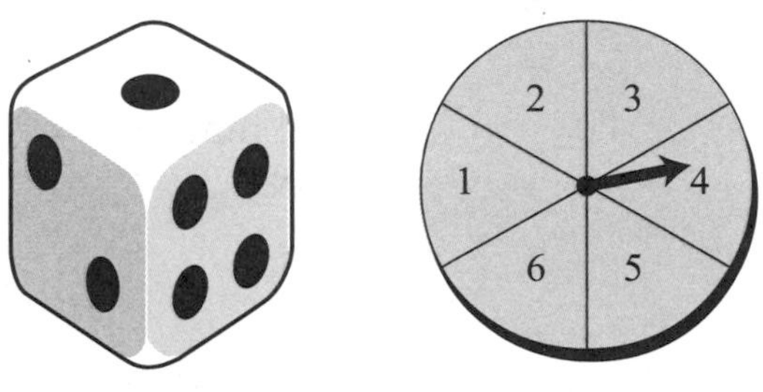

These events are independent events. The outcome of one event *does not* affect the outcome of the other event.

In this example, what you roll on the die has no affect on what you spin on the spinner.

$$P(4) = \frac{1}{6} \quad P(\text{even number on spinner}) = \frac{3}{6} \text{ or } \frac{1}{2}$$

Example

There are two dimes, a quarter, and four nickels in a bag. Once a coin is chosen, it is *not* replaced. What is the probability of choosing a nickel and then a dime?

This is an example of dependent events. The outcome of one event *does* affect the outcome of the other event.

$$P(\text{nickel}) = \frac{4}{7} \quad P(\text{getting a dime second}) = \frac{2}{6} = \frac{1}{3}$$

Notice there are only 6 coins left, since the first coin drawn was not replaced.

RULE: To find the probability of two dependent events A and B, multiply the probability of event A by the probability of event B after event A has occurred.

Examples

1. There are 4 red, 5 white, and 3 blue marbles in a bag. Find the probability of selecting a red marble and then a white marble, if once a marble is chosen it is *not* replaced.

Solution: P (1st marble is red) $= \frac{4}{12}$ ← Number of red marbles / ← Total number of marbles

P (2nd marble is white) $= \frac{5}{11}$ ← Number of white marbles / ← Total number of marbles after the 1st red marble is drawn

P (1st marble is red and 2nd marble is white) $= \frac{4}{12} \cdot \frac{5}{11}$

$= \frac{20}{132}$

$= \frac{5}{33}$

2. Katrina wrote each letter of the word MISSISSIPPI on cards and put them into a hat. She pulled out a card, and without replacing it, drew a second card. What is the probability that both cards Katrina picked have an S on them?

Solution: P (first card with an S on it) $= \frac{4}{10}$ ← Number of cards with S on it / ← Total number of cards

P (second card with an S on it) $= \frac{3}{9}$ ← Number of cards with S on it after the first card is drawn / ← Total number of cards after the first card is drawn

P (both cards have an S on it) $= \frac{4}{10} \cdot \frac{3}{9}$

$= \frac{12}{90}$

$= \frac{2}{15}$

TEST YOUR SKILLS (For answers, see page 276.)

1. Anthony tossed a coin and rolled a cube numbered 1 through 6. What is the probability that he will get a heads and roll an even number?

   A. 0.25

   B. 0.5

   C. 0.75

   D. 1.0

2. There are 15 girls and 8 boys in Ms. Lopez's math class. Each student wrote their name on a slip of paper and placed it in a hat. Two names were drawn from the hat for a free homework pass. What is the probability that the first name chosen was a girl and the second name chosen was a boy?

   A. $\frac{30}{59}$

   B. $\frac{60}{253}$

   C. $\frac{3}{10}$

   D. $\frac{2}{95}$

3. Suppose you pick two cards from a standard deck of 52 cards. Once a card is drawn, it is *not* replaced. What is the probability that both cards drawn are red?

   A. $\frac{13}{52} \cdot \frac{12}{51}$

   B. $\frac{13}{52} \cdot \frac{13}{52}$

   C. $\frac{26}{52} \cdot \frac{25}{51}$

   D. $\frac{26}{52} \cdot \frac{26}{52}$

4. There are 5 lollipops, 4 pieces of gum, and 12 taffies in a bag of candy. Once a piece of candy is selected, it is *not* replaced. What is the probability that a piece of gum is chosen and then a lollipop?

A. $\frac{4}{7}$

B. $\frac{1}{28}$

C. $\frac{1}{21}$

D. $\frac{9}{41}$

5. There are 5 white socks, 6 brown socks, and 4 black socks in a drawer. Once a sock is drawn it is *not* replaced. What is the probability that both socks chosen are *not* brown?

A. $\frac{9}{25}$

B. $\frac{12}{35}$

C. $\frac{3}{5}$

D. $\frac{4}{15}$

6. Melissa wrote each letter of the word ALGEBRA on a card and put them into a hat. What is the probability that the first card drawn is a vowel and the second card drawn is the letter G, if the first card is *not* replaced?

A. $\frac{1}{14}$

B. $\frac{2}{21}$

C. $\frac{3}{13}$

D. $\frac{3}{14}$

7. You have a basket of fruit. The number of different fruits is shown in the table below. Without replacing the first fruit, you choose another. What is the probability that you choose an apple and then an orange?

Fruit	Number of Fruits
Apples	10
Bananas	4
Oranges	6

A. $\frac{1}{60}$

B. $\frac{3}{20}$

C. $\frac{4}{5}$

D. $\frac{3}{19}$

8. There are 6 red, 8 blue, 4 yellow, and 2 green marbles in a bag. One marble is drawn, *not* replaced, and then a second marble is drawn. What is the probability that the first marble is yellow and the second marble is green?

A. $\frac{1}{380}$

B. $\frac{2}{95}$

C. $\frac{2}{13}$

D. $\frac{1}{50}$

Short Response

9. The letters of the words GREATEST COMMON FACTOR are written on cards and thrown into a hat. A card is picked, *not* replaced, and then a second card is chosen.

Part A.

Find the probability that neither card is a vowel. **Show your work.**

Answer __________

Part B.

Does the probability that a vowel is drawn second depend on the first outcome? **Explain your answer.**

____

____

____

10. James had to solve the following problem for homework. A deck of cards numbered 0–9 are put into a hat. A card is drawn, *not* replaced, and then a second card is chosen. His work is shown below.

P (first number even) $= \frac{5}{9}$

Answer: $\frac{5}{9} \cdot \frac{4}{9} = \frac{20}{81}$

P (second number even) $= \frac{4}{9}$

Part A.

Describe the two errors James made on his homework when he tried to find the probability of choosing two even numbers.

____

____

____

Part B.

Show James how to correctly solve the problem. Show your work.

Answer __________

EXPERIMENTAL PROBABILITY

Laurie flipped a coin 20 times. Her results are shown in the table below.

H	T	H	H	H	T	T	H	H	H
H	H	H	T	H	T	H	T	H	H

After the experiment, Laurie reviewed her results. Out of 20 tosses, she got 14 heads and 6 tails.

Laurie's experimental probability of getting a head was $\frac{14}{20}$.

Experimental probability—probabilities based on the outcomes that you obtain during an experiment. These can change each time you run the experiment. For example, if Laurie ran this experiment again, next time she might only get 8 heads out of 20.

Laurie's theoretical probability of getting a head is $\frac{1}{2}$.

Theoretical probability—probabilities based on known facts. For example, the probability of getting a head anytime you flip a coin is one out of two.

EXAMPLES

1. As an experiment, you flipped a coin 25 times and got 15 tails. What is the experimental probability of getting a tails?

   Solution: The experimental probability is what you got by doing the experiment, $\frac{15}{25} = \frac{3}{5}$.

2. At a local middle school, 125 students out of 500 said they own a cell phone. What is the probability that a student does *not* own a cell phone?

Solution: If $\frac{125}{500}$ students own a cell phone, then 500 – 125 equals the number of students who do not own a cell phone.

$500 - 125 = 375$

The probability is $\frac{375}{500} = \frac{3}{4}$.

3. Out of 500 people, how many would you expect to list rock as their favorite music?

Favorite Music	
Country	10
Rock	25
Classical	5
Rap	10

Solution:

Find the experimental probability of rock music being the favorite.

$$P\text{ (rock)} = \frac{25}{50} = \frac{1}{2}$$

Set up a proportion to see how many people out of 500 like rock music.

$$\frac{1}{2} = \frac{x}{500}$$

$$\frac{2x}{2} = \frac{500}{2}$$

$$x = 250$$

You could expect 250 people to list rock as their favorite music.

TEST YOUR SKILLS (For answers, see page 277.)

1. At a recent basketball game, Joey made 6 out of 20 shots. What is the probability that he will make the next shot?

   A. $\frac{1}{2}$

   B. $\frac{3}{10}$

   C. $\frac{7}{10}$

   D. $\frac{1}{5}$

2. Thirteen percent of the seventh grade boys ride snowmobiles. If there are 200 boys in the seventh grade, how many boys ride snowmobiles?

   A. 13
   B. 26
   C. 32
   D. 52

For questions 3 and 4, use the table below.

Favorite Television Show	
Name of Show	Number of Students Who Watch
Lost	25
American Idol	42
The Office	13
CSI	20

3. Students at a local middle school were asked to name their favorite TV show. Based on the table, what is the probability that a student watched the show *Lost*?

A. $\frac{1}{4}$

B. $\frac{1}{2}$

C. $\frac{2}{3}$

D. $\frac{2}{5}$

4. If 500 students were surveyed, what is a good prediction of the number of students who will watch *CSI*?

A. 40

B. 60

C. 80

D. 100

For questions 5–7, use the table below that shows the results of spinning a spinner 40 times with equal sectors labeled 1, 2, 3, 4, and 5.

Result	Frequency
1	10
2	9
3	7
4	8
5	6

5. Based on the results, what is the probability of spinning a 3?

A. $\frac{1}{5}$

B. $\frac{7}{40}$

C. $\frac{1}{10}$

D. $\frac{3}{20}$

6. What is the theoretical probability of spinning a 3?

A. $\frac{1}{2}$

B. $\frac{1}{5}$

C. $\frac{7}{40}$

D. $\frac{1}{10}$

7. Out of 400 spins, what is a good prediction of the number of times you would get a 5?

A. 40

B. 50

C. 60

D. 70

8. Juan made 15 out of 20 hits at the target during archery practice. What is a good prediction of the number of hits Juan will make if he shoots 300 times?

A. 150

B. 200

C. 225

D. 275

9. One out of four students in the Armstrong Middle School said they own an iPod. If 500 students are surveyed, how many students do *not* own an iPod?

A. 275

B. 300

C. 325

D. 375

Short Response

10. Mr. Neal ran an experiment in his math class. He had five of his students flip a coin 20 times and record the results. The results are listed in the table below.

Student Name	Heads	Tails
Mike	11	9
Kevin	15	5
Lisa	8	12
Amy	6	14

Part A.

According to the experimental probabilities, who is most likely to get a tails on the next toss?

Answer __________

Part B.

If Lisa were to flip a coin 80 times, what is a good prediction of the number of heads she would get? **Show your work.**

Answer __________

Chapter 6

STATISTICS

FREQUENCY TABLES

A frequency table is a table or chart that shows how often data occurs. When data is organized, it is easier to read and analyze.

Examples

1. Which frequency table shows a cumulative frequency of 20 students who participated in a school activity?

A.

Activity	Frequency
School Musical	10
Yearbook Club	5
Newspaper Club	4
Chess Club	3

B.

Activity	Frequency
School Musical	8
Yearbook Club	5
Newspaper Club	7
Chess Club	11

C.

Activity	Frequency
School Musical	9
Yearbook Club	5
Newspaper Club	4
Chess Club	3

D.

Activity	Frequency
School Musical	6
Yearbook Club	5
Newspaper Club	6
Chess Club	3

To find the cumulative frequency, add the frequency for each activity to find the total number of students. Choice A adds to 22, B adds to 31, C totals 21, and D totals 20. Choice D has a cumulative frequency of 20 students.

2. Complete the frequency table below for the letters in the word MISSISSIPPI.

Letter	Tally	Frequency

Fill in the table with the different letters in the word MISSISSIPPI and then count how many times each of the letters appears. The completed table should look like the following:

Letter	Tally	Frequency
M	|	1
I	||||	4
S	||||	4
P	||	2

TEST YOUR SKILLS (For answers, see page 279.)

For questions 1–3, use the frequency table below.

FAVORITE SPORT

Favorite Sport	Number of Students
Football	10
Basketball	5
Baseball	2
Soccer	3

1. How many students were surveyed?

   A. 5
   B. 8
   C. 15
   D. 20

2. What percent of the votes were for baseball?

   A. 2%
   B. 5%
   C. 10%
   D. 20%

3. Which two sports together received 75% of the votes?

   A. baseball and soccer
   B. football and basketball
   C. basketball and baseball
   D. football and soccer

For questions 4–7, use the following information.

In a bag of Skittles, 30% of the Skittles arc brown, 20% blue, 10% green, 10% orange, and 30% yellow.

4. Complete the frequency table.

Color of Skittle	Percent

5. If there are 30 Skittles in the bag, how many Skittles are green?

A. 1
B. 2
C. 3
D. 4

6. Which two colors make up 50% of the Skittles?

A. orange and yellow
B. orange and blue
C. brown and blue
D. blue and green

7. The frequency table shows all of the following are true statements except which one?

A. The highest percentage of Skittles are brown.
B. Thirty percent of the Skittles are yellow.
C. People wish that more of the Skittles were blue.
D. Twenty percent of the Skittles are green and orange.

For questions 8–10, use the following information. The table below lists the number of pets people own.

1	1	2	3	1	0
0	0	2	3	1	1
1	2	1	1	4	0

8. Complete the frequency table below using the information from the table above.

Number of Pets	Tally	Frequency
0		
1		
2		
3		
4		

9. How many people were surveyed?

A. 10
B. 12
C. 18
D. 20

10. What is the most common number of pets that people own?

A. 0
B. 1
C. 2
D. 3

Short Response

11. Reagan and her friends collect marbles. Reagan has 5 red, 3 blue, and 4 white marbles. Jack has 8 red, 2 blue, and 8 white marbles. Josey has 2 red, 11 blue, and 1 white marble.

Part A.

Draw a frequency table showing the total number of marbles for each color.

Part B.

When combined, it seems that the white marble is the most popular. Do you agree? Explain your answer.

____

____

____

VENN DIAGRAMS

A Venn diagram is a pictorial way of showing relationships between sets. Circles are used to represent sets. The area where the circles overlap shows the data that the sets have in common.

Example

Fourteen students were asked which winter activity they preferred. The results are shown in the Venn diagram below.

7th Grade Winter Activities

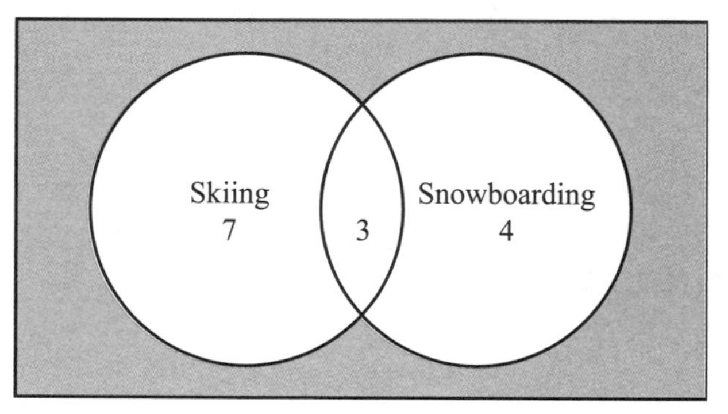

The following information can be obtained using the Venn diagram above.

- 10 students ski
- 7 students snowboard
- 7 students *only* ski

- 4 students *only* snowboard
- 3 students ski *and* snowboard
- 14 students ski *or* snowboard

Example

Students were asked to name the sport or sports that they were involved in during the school year. The results are shown below.

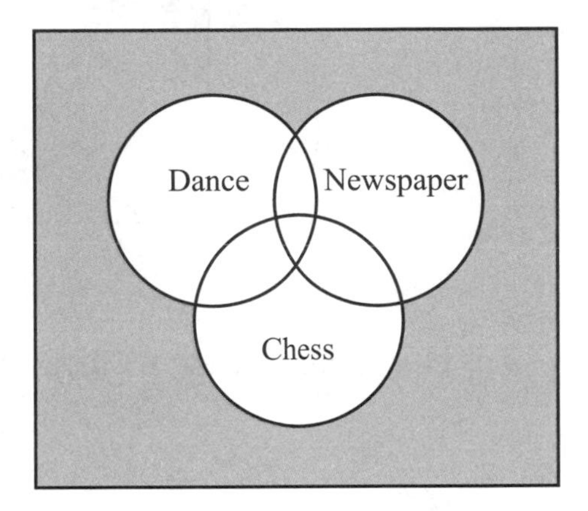

Which statement is true?

A. All students in the newspaper club are also in the chess club.

B. Some students are in dance and the newspaper club.

C. Some students are in dance and the archery club.

D. None of the chess players are in another activity.

Solution:

Choice B is a true statement.
Choice A is false because not all of the students are in the chess club. There are also some in dance.
Choice C is false because the Venn diagram does not show an archery club.
Choice D is false because the chess club overlaps with dance and the newspaper club, therefore some students are also in these activities.

TEST YOUR SKILLS (For answers, see page 281.)

Use the Venn diagram to answer questions 1–5.

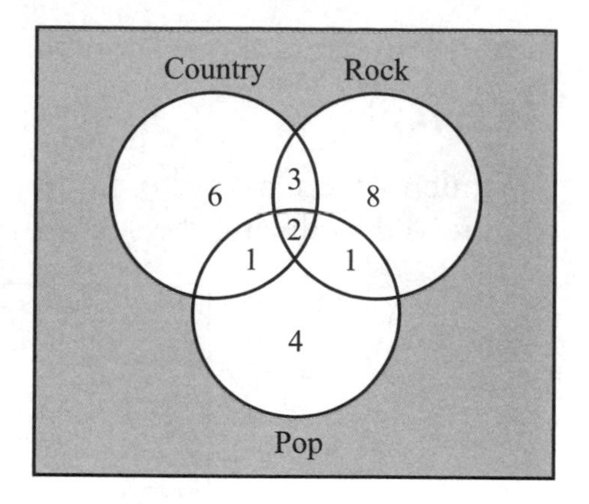

1. How many students listen to country music?

   A. 3
   B. 6
   C. 8
   D. 12

2. How many students listen only to rock music?

   A. 1
   B. 2
   C. 3
   D. 8

3. How many students listen to all three types of music?

   A. 1
   B. 2
   C. 3
   D. 7

4. How many students listen to country and pop music?

   A. 1
   B. 4
   C. 6
   D. 8

5. If 30 students were surveyed, how many students listen to some type of music other than country, rock, or pop?

   A. 4
   B. 5
   C. 7
   D. 8

Your school just had seventh-grade gym sign-ups. There are 30 students in fourth period PE class. Fourteen students signed up for volleyball, 16 students signed up for basketball, and 2 students signed up for both. Use this information to help you answer questions 6–9.

6. Using the information above, fill in the Venn diagram.

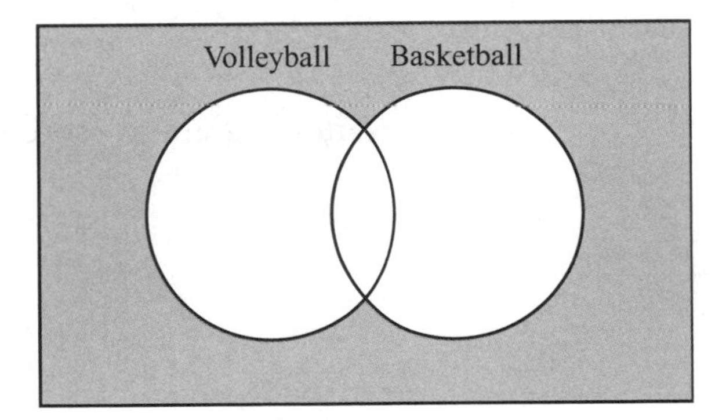

7. How many students signed up for volleyball only?

   A. 2
   B. 12
   C. 14
   D. 26

8. How many students like basketball only?

A. 2
B. 12
C. 14
D. 16

9. How many students must have signed up for another sport besides volleyball or basketball?

A. 2
B. 5
C. 7
D. 10

Short Response

10. In a survey of 40 people, every person had been to either a Chicago Bulls game, a New York Knicks game, or both. Eighteen people had been to a Chicago Bulls game and eight people had been to a Chicago Bulls and a New York Knicks game.

Part A.

Draw a Venn diagram to represent the situation stated above.

Part B.

How many people have been to a New York Knicks game only?

Answer __________

COLLECTING, READING, AND INTERPRETING DATA

A graph is a way to display and compare data.

Type of Graph	Definition	Example
Pictograph	A graph that uses a picture to represent how many times an object occurs.	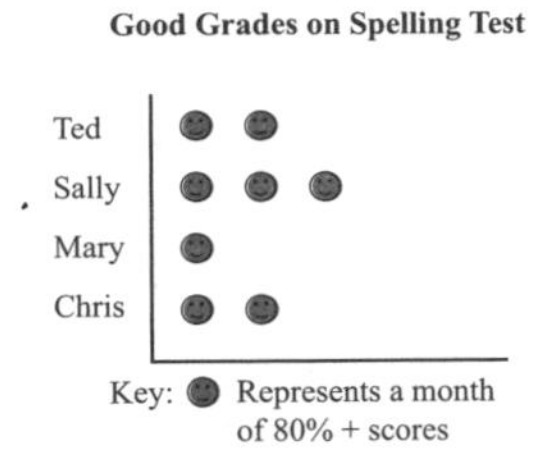
Bar Graph	A graph that uses bars of different lengths to compare items or objects.	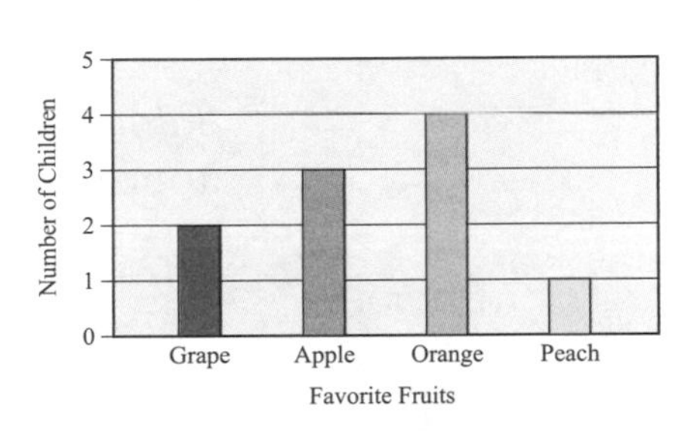
Histogram	A type of bar graph that shows how often numerical data occurs in equal intervals.	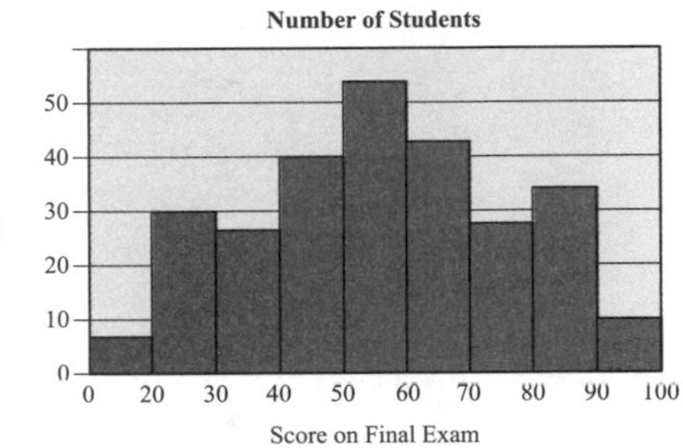
Line Graph	A graph that displays data over a period of time.	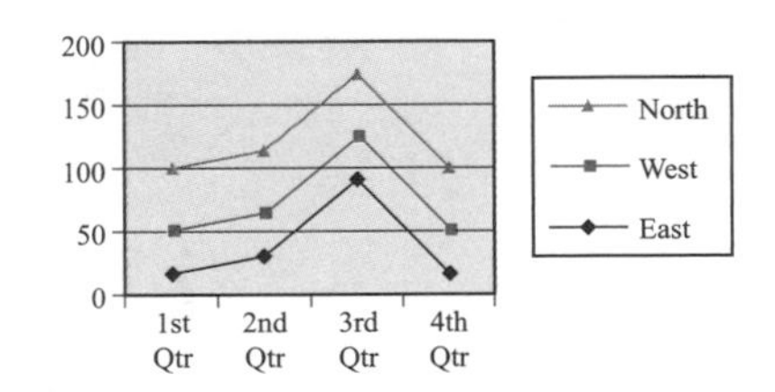
Circle Graph	A graph that compares data as parts of a whole.	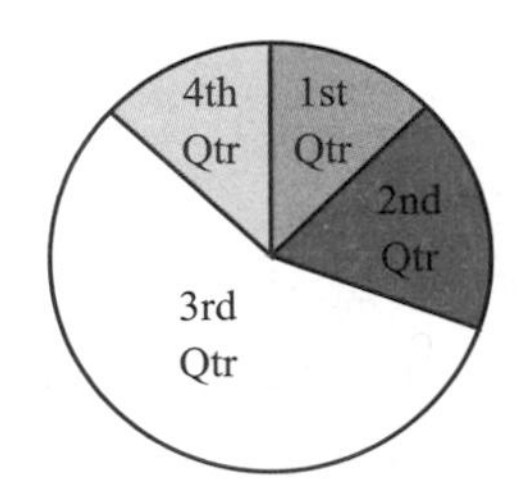

Example

The table below shows the results of a survey that asked students who they voted for as class president.

Candidate	Percent
Rob	32%
Julie	38%
Catherine	30%

Which graph could best represent the data?

Solution: The percents add up to 100%, so a circle graph would be appropriate.

Examples

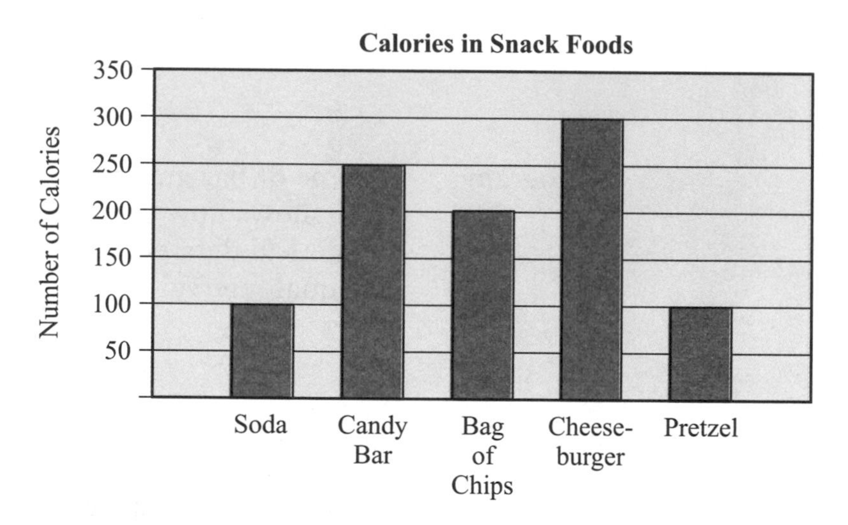

1. Which item has the most calories?

   Solution: The cheeseburger, since that bar is the highest.

2. How many more calories does a candy bar have than a pretzel?

   Solution: A candy bar has 250 calories. A pretzel has 100 calories.

   $250 - 100 = 150$

   The candy bar has 150 more calories than a pretzel.

TEST YOUR SKILLS (For answers, see page 283.)

1. The boys' soccer team recorded the number of candy bars they sold on each day of the week. Which graph could best represent the data?

   A. pictograph
   B. circle graph
   C. bar graph
   D. histogram

2. Ms. Stahl wanted to see how many bonus points her students received during the year from 0–2, 3–5, 6–8, and 9–11. Which type of graph would best represent the data?

   A. bar graph
   B. circle graph
   C. pictograph
   D. histogram

3. The price of gas has been recorded for the months of April, May, June, July, and August. Which type of graph would be best to show whether the price of gas has increased or decreased over these few months?

   A. bar graph
   B. pictograph
   C. line graph
   D. histogram

Use the pictograph below to answer questions 4 and 5.

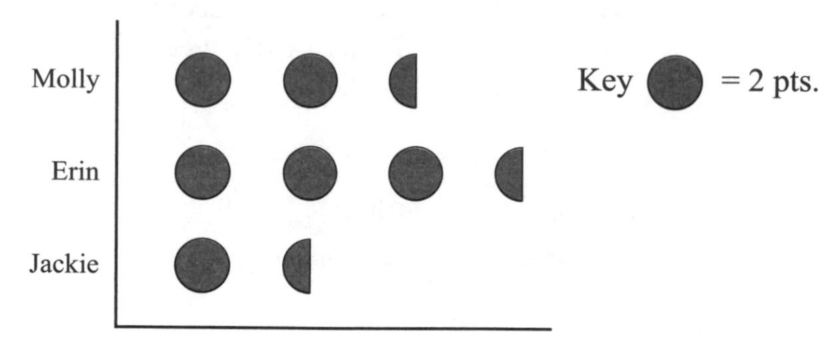

4. How many bonus points did Molly earn?

A. 2.5
B. 5
C. 6
D. 6.5

5. How many bonus points were given out in all?

A. 8.5
B. 12
C. 12.5
D. 13.5

Use the line graph below for questions 6 and 7.

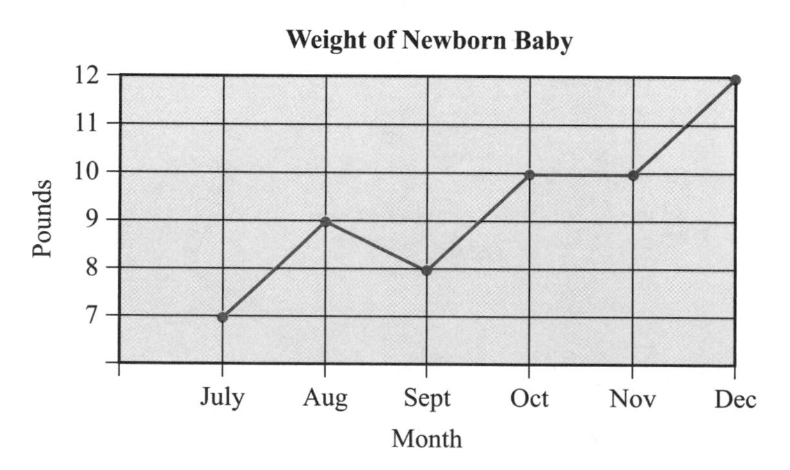

6. How much did the baby weigh in August?

A. 7 lbs.

B. 8 lbs.

C. 9 lbs.

D. 10 lbs.

7. What was the difference between the baby's weight in September and in December?

A. 3 lbs.

B. 4 lbs.

C. 5 lbs.

D. 6 lbs.

The histogram below shows retirement ages of people in Williams School District. Use the histogram to answer questions 8–10.

8. How many more people retired between the ages of 60–64 than between the ages of 50–54?

A. 25

B. 30

C. 35

D. 40

9. How many people retired between 55 and 69 years of age?

A. 90
B. 100
C. 110
D. 120

10. How many people were at least 65 years old when they retire?

A. 35
B. 45
C. 50
D. 55

Short Response

11. The eighth graders at Armstrong Middle School can choose between four different field trips. The results of the survey are shown below.

Field Trip Activity	Number of Students in Favor
A Picnic	20
A Formal Dance	35
A Field Day	15
Darien Lake Day	45

Part A.

Which type of graph could be used to best display the data?

Answer ___________

Part B.

Explain your answer.

____

____

____

RANGE

The **range** of data is the difference between the highest value and the lowest value.

Example

The data listed below shows how many minutes students in Mr. Lee's math class study for a quiz.

12, 10, 15, 16, 20, 5, 45, 25, 30, 60

What is the range of this data?

Solution:

The highest minutes are 60 and the lowest are 5. So $60 - 5 = 55$.

Example

The following are the temperatures recorded during a week in February.

10, 8, –2, –6, 0, 5, 12

What is the range of the temperatures?

Solution:

The highest temperature is 12 and the lowest temperature is –6.

$12 - (-6)$
$12 + 6$
18

The range of the temperature is 18.

TEST YOUR SKILLS (For answers, see page 283.)

1. These are the prices of six different student lunches bought on the same day. What is the range of these prices?

$6.40, $3.15, $2.25, $5.05, $3.75, $2.10

A. $2.25
B. $3.75
C. $4.30
D. $8.50

2. Below is the data showing how many text messages students send on an average day. What is the range of this data?

12	15
10	25
5	45
60	20
35	75
30	40

A. 5
B. 55
C. 70
D. 75

3. Sara's test scores are shown below. What is the range of her test scores?

97, 85, 83, 95, 84, 72, 68, 88, 99

A. 28
B. 31
C. 84
D. 99

4. Scott scored the second highest grade in his math class by 8 points. The range of the test scores was 25 points. If the lowest test score was a 62, what was Scott's grade?

A. 93
B. 79
C. 81
D. 95

5. The following temperatures were recorded in Buffalo for seven days.

$$13°, 5°, 23°, -4°, 28°, 15°, -6°$$

What is the range of these temperatures?

A. 22°

B. 24°

C. 30°

D. 34°

6. Kevin beat the lowest score on a test by 2 points. The range on the test was 42 points. If the highest score was a 92, what did Kevin get on the test?

A. 48

B. 50

C. 52

D. 56

7. The following table lists the number of students who participated in the school musical each year.

2002	2003	2004	2005	2006	2007
49	63	93	102	86	79

What is the range of the data?

A. 23

B. 39

C. 50

D. 53

8. The line graph below shows the number of hours it rained during a week. Use the line graph to answer the following question.

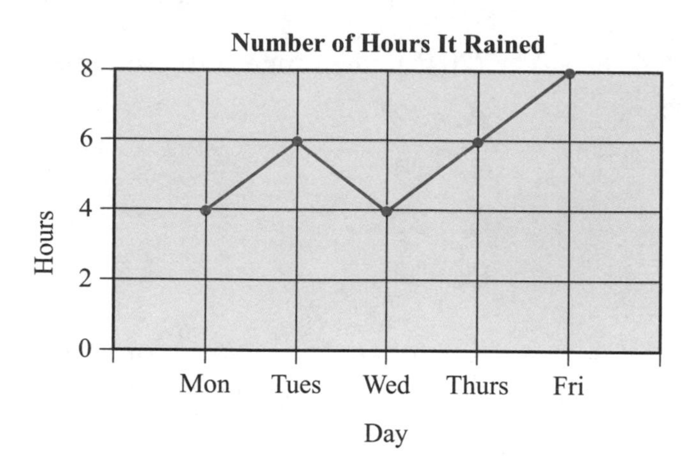

What is the range of the number of hours it rained?

A. 2

B. 3

C. 4

D. 8

Short Response

9. A local ski company took a survey to see how many times people skied during the ski season.

NUMBER OF TIMES SKIED DURING THE SEASON

9	16	25	52	6	3	48	33
21	18	11	4	12	36	42	16

Part A.

What is the range of the number of times people skied?

Answer __________

Part B.

Explain using words and numbers why your answer is correct.

____

____

____

MEASURES OF CENTRAL TENDENCY

Three averages can be used to describe the center of a data set. These values are known as the measures of central tendency.

THE MEASURES OF CENTRAL TENDENCY INCLUDE:

Mean The sum of given numbers divided by how many values there are total. Also known as the numerical average.

Median The middle number when a set of numbers are in number order.

- If the set has an odd number of data, the median is the middle number.
- If the set has an even number of data, the median is the average of the two middle numbers.

Mode The number or numbers that occur most often in a set of data.

You will use these averages to describe a set of data. You will also have to choose which measure(s) best describes the data.

Example

These are the salaries of five people who work at Northside Middle School. Which measure(s) of central tendency best describes the data?

28,000, 32,000, 34,500, 45,000, 75,000

A. mode

B. mean and median

C. median and mode

D. range

Solution:

Find the mean, median, and mode. Since range is not a measure of central tendency, choice D can be eliminated.

$$\text{Mean} = \frac{28{,}000 + 32{,}000 + 34{,}500 + 45{,}000 + 75{,}000}{5}$$

$$\frac{214{,}500}{5}$$

$$\$42{,}900$$

Median = 28,000, 32,000, 34,500, 45,000, 75,000

When the numbers are in numerical order, the middle number is 34,500.

Mode = Since there is not a number that occurs most often, there is no mode.

Since the mean and the median both are in the center of all of the data, choice B is the correct answer.

TEST YOUR SKILLS (For answers, see page 284.)

1. The following averages represent the rainfall in inches of eight cities. Find the mean to the nearest tenth.

3.2, 4.1, 6.0, 0.4, 1.2, 1.8, 0.9, 1.8

A. 0.8

B. 1.8

C. 2.4

D. 19.4

2. The following data represents the heights of students in inches on the Armstrong Middle School's boys basketball team. Find the median height.

58, 55, 52, 60, 53, 56, 61, 60

A. 55

B. 57

C. 60

D. 61

3. Using the following numbers, which is a true statement?

16, 28, 10, 11, 23

A. The mean is 45.

B. The median is 10.

C. The mode is 11.

D. There is no mode.

4. The data set below shows Abib's test scores. Which measure(s) of central tendency best describes the data?

42, 89, 92, 85, 88, 74, 63

A. median

B. mean and mode

C. median and mode

D. mean and median

5. Which set of data has a median of 51?

A. 48, 50, 52, 56

B. 49, 50, 52, 53, 54

C. 50, 50, 50, 51, 52

D. 51, 52, 53, 54

6. Which measure of central tendency best describes the data below?

0, 84, 86, 90, 92, 92

A. mean
B. median
C. mode
D. range

7. Using the following set of data, which statement is true?

20, 22, 58, 60, 100, 100

A. mean < median
B. mean > mode
C. mode > median
D. median > mean

8. Which of the following is *not* a true statement about the data below?

18, 18, 20, 24

A. mean > mode
B. median > mode
C. median < mean
D. mode > mean

Short Response

9. The average wait times in minutes for eight different rides at an amusement park are 44, 37, 22, 11, 17, 25, 34, and 17 minutes.

Part A.

Which measure of central tendency would the amusement park use to encourage people to come to the park? Show your work.

Answer ________

Part B.

Using what you know about measures of central tendencies, explain why your answer in Part A is correct.

____

____

____

CIRCLE GRAPHS

A circle graph compares parts to the whole. The whole is represented as the circle, and the parts are represented as the sections of the circle.

Example

Bobby surveyed his seventh grade math class to find out everyone's favorite sport. His results are displayed in the table below.

FAVORITE SPORT

Football	35%
Baseball	45%
Hockey	20%

Make a circle graph using the information given.

Step 1

There are two methods you can use to determine how many degrees out of 360 each percent represents. Both methods are shown.

Method 1—Chart Method

There are 360° in a circle. Multiply each percent by 360 to find the number of degrees in each section of the circle. You can use this chart every time to help you organize your data.

Percent	Decimal (360)	Degrees
35%	.35(360)	126°
45%	.45(360)	162°
20%	.20(360)	72°

Method 2—Proportion Method

You can also set up a proportion to find the number of degrees each sport would represent in the circle.

Football

$$\frac{35}{100} = \frac{x}{360}$$
$$35(360) = 100x$$
$$\frac{12{,}600}{100} = \frac{100x}{100}$$
$$x = 126°$$

Baseball

$$\frac{45}{100} = \frac{x}{360}$$
$$45(360) = 100x$$
$$\frac{16{,}200}{100} = \frac{100x}{100}$$
$$x = 162°$$

Hockey

$$\frac{20}{100} = \frac{x}{360}$$
$$20(360) = 100x$$
$$\frac{7200}{100} = \frac{100x}{100}$$
$$x = 72°$$

Step 2

Use a compass to draw a circle and a radius. Use a protractor to draw a 126° angle. From the new radius, draw the next angle of 162°. From the new radius, measure the last section of the circle. This should equal the last angle of 72°.

Step 3

Label each section of the graph with the name of each sport and the percent. Also give the graph a title.

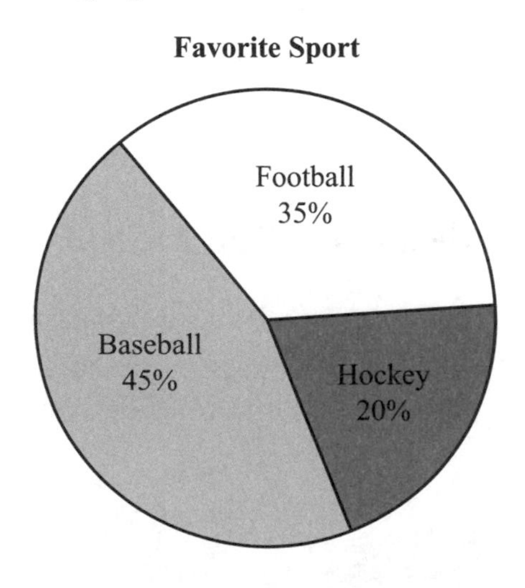

Example

A group of middle school students were asked to name their favorite food. The results are shown in the table below.

FAVORITE FOOD

Favorite Food	Number of Students
Pizza	8
Subs	2
Tacos	6
Hamburgers	4

Use your protractor to make a circle graph of the data.

Solution:

Using the table method:

- Find the total number of students surveyed.
- Find the ratio that compares the number of students who like each food to the total number of students.
- Change each ratio to a percent (decimal) and multiply this by 360 to find the degrees.

Ratio	Decimal (360)	Degrees
8/20	0.4 (360)	144°
2/20	0.1 (360)	36°
6/20	0.3 (360)	108°
4/20	0.2 (360)	72°

- Use a compass to draw a circle and a radius. Use a protractor to draw a 144° angle. From the new radius, draw the next angle of 36°. From the new radius, draw the next angle of 108°. From the new radius, measure the last section of the circle. This should equal the last angle of 72°.

- Label each section of the graph with the name of the favorite food and the percent. Also give the graph a title.

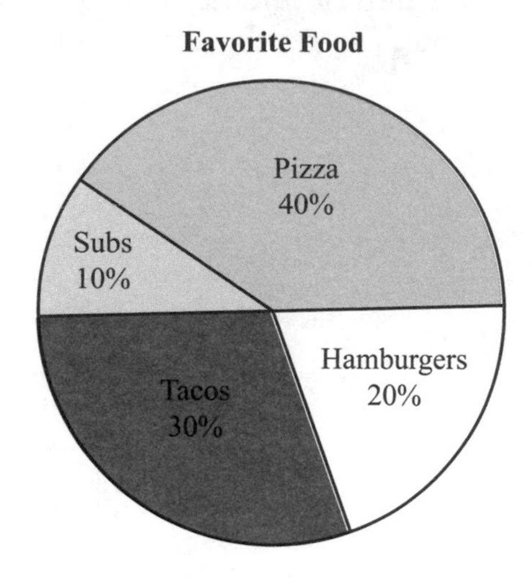

TEST YOUR SKILLS (For answers, see page 285.)

1. What central angle represents 50%?

   A. 50°
   B. 90°
   C. 100°
   D. 180°

2. What central angle represents 30%?

   A. 30°
   B. 60°
   C. 108°
   D. 127°

3. If the central angle measures 72°, what percent of the whole circle does this central angle represent?

   A. 15%
   B. 20%
   C. 25%
   D. 30%

4. If the central angle measures 126°, what percent of the whole circle does this central angle represent?

   A. 25%

   B. 30%

   C. 35%

   D. 40%

Use the circle graph below to answer questions 5 and 6.

Brianna surveyed the students in the seventh grade at her school to find out which summer activity was their favorite. The results are shown in the circle graph below.

FAVORITE ACTIVITIES

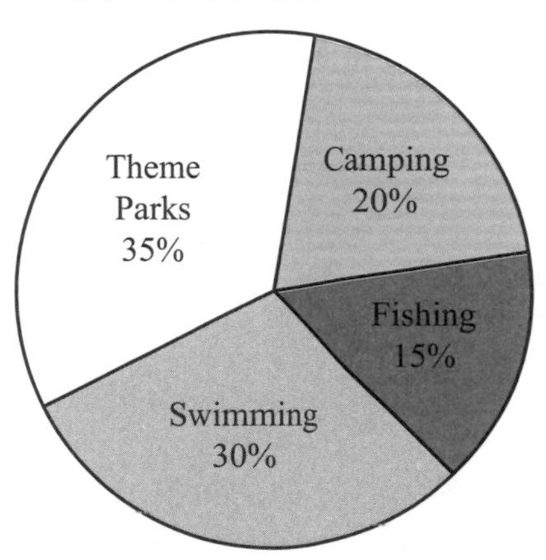

5. What is the measure of the central angle representing the percentage of students who like to fish in the summer?

   A. 5.4°

   B. 15°

   C. 54°

   D. 100°

6. What is the measure of the central angle representing the percentage of students who like to go to theme parks in the summer?

   A. 35°
   B. 112°
   C. 126°
   D. 180°

Use the table below to answer questions 7 and 8.

A local neighborhood was surveyed about what type of pets they owned. The results are shown in the table below.

Number of Families
Who Own Pets

Hamsters	3
Cats	8
Dogs	6
None	3

7. What is the measure of the central angle representing the number of people who own a dog?

   A. 35°
   B. 90°
   C. 100°
   D. 108°

8. If you made a circle graph of families who own pets, which pet would have a central angle of 96°?

   A. Hamsters
   B. Cats
   C. Dogs
   D. None

Short Response

9. The table below shows gifts that students want for their birthday.

GIFTS STUDENTS WANT

Item	Number of Students
iPod	6
Cell phone	12
PlayStation	3
Computer	9

Part A.

Using your protractor, make a circle graph of the data given above.

Be sure to include:

- All of your work
- Labeled sections for each part of the circle
- A title for your circle graph

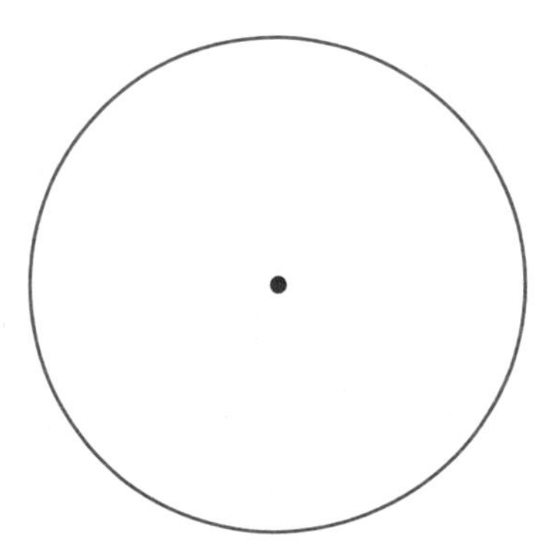

Part B.

What percent of the students want an iPod for their birthday?

Answer __________

DOUBLE BAR GRAPHS

A double bar graph is a bar graph that that uses two bars to make comparisons.

Examples

The double bar graph below shows car sales from 2000 to 2002 for Ford and Toyota.

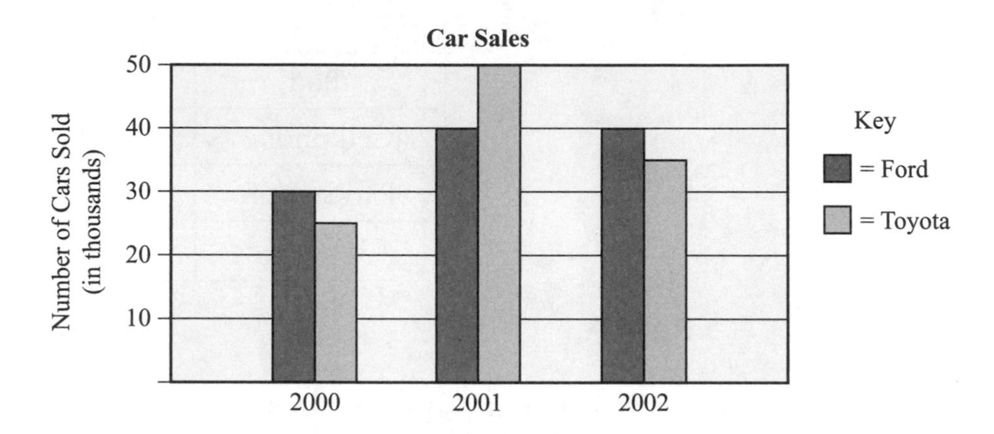

1. How many more Fords sold in the year 2000 than Toyotas?

Solution:

Ford sold 30,000 cars in 2000 and Toyota sold 25,000 cars. Ford sold 5,000 more cars in 2000 than Toyota.

2. What conclusion can you make based on the graph?

A. More Toyotas sold than Ford from 2000 to 2002.

B. Toyota sold 5000 more cars than Ford in 2000.

C. Toyota sold 10,000 more cars than Ford in 2001.

D. More people like Ford than Toyota.

Solution:

C In 2001, Toyota sold 50,000 cars and Ford sold 40,000 cars.

TEST YOUR SKILLS (For answers, see page 287.)

The double bar graph below compares the number of minutes Sam and John spend on the Internet each day of the week. Use the double bar graph to answer questions 1–3.

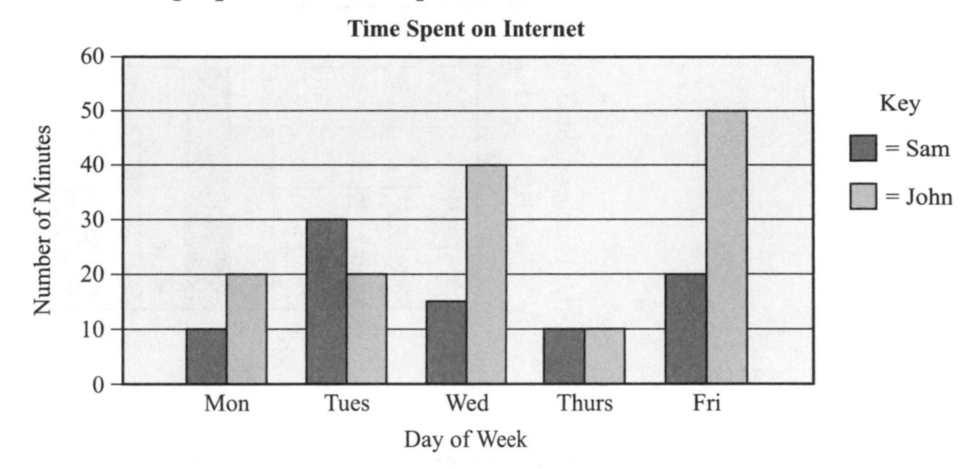

1. How many more minutes did Sam spend on the Internet on Tuesday than John?

   A. 10

   B. 15

   C. 20

   D. 30

2. How many more minutes did John spend on the Internet during the week than Sam?

   A. 35

   B. 55

   C. 70

   D. 75

3. On which day did Sam and John spend the same amount of time on the Internet?

   A. Monday

   B. Tuesday

   C. Thursday

   D. Friday

The double bar graph below compares the number of students in sixth, seventh, and eighth grades that buy school lunch. Use the double bar graph to answer questions 4–6.

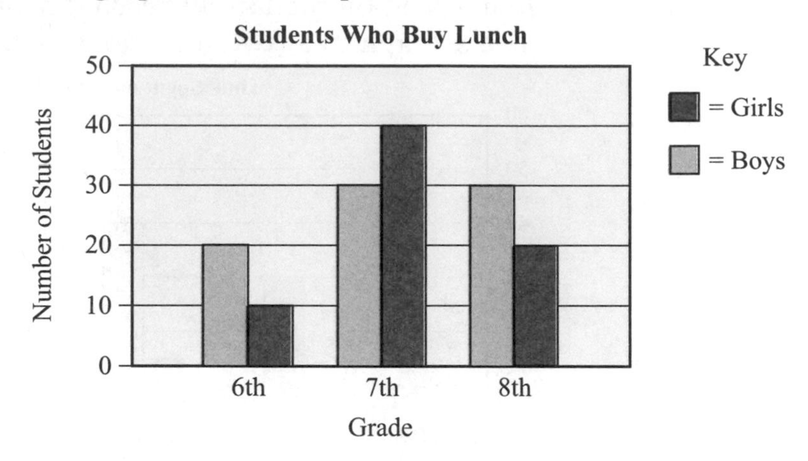

4. In sixth, seventh, and eighth grades combined, how many more boys buy lunch than girls?

A. 5

B. 10

C. 15

D. 20

5. How many more seventh-grade girls buy lunch than sixth-grade girls?

A. 20

B. 30

C. 35

D. 40

6. Which conclusion can you make based on the graph?

A. More sixth-grade girls buy lunch than sixth-grade boys.

B. More seventh-grade boys buy lunch than seventh-grade girls.

C. Twenty more seventh-grade girls buy lunch than eighth-grade girls.

D. In every grade level, boys buy lunch more than girls.

The double bar graph below shows the number of students who graduated from 2003 to 2007. Use the double bar graph to answer questions 7–10.

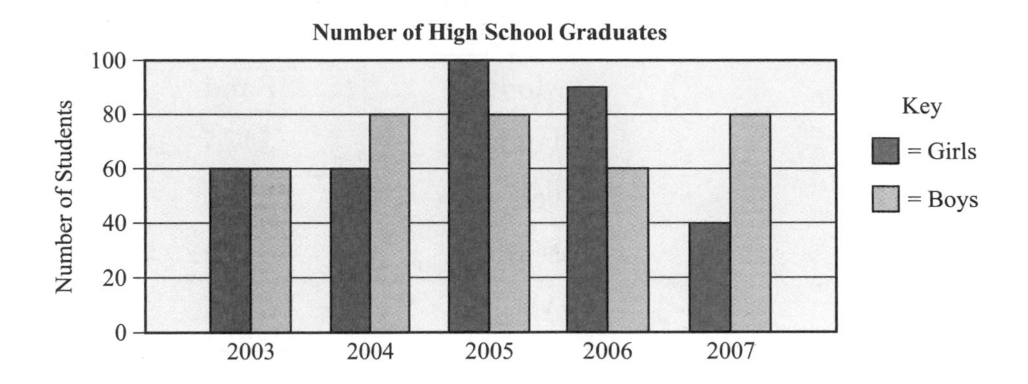

7. Which year had 20 more boys graduate than girls?

A. 2003

B. 2004

C. 2005

D. 2006

8. How many more girls graduated in 2006 than in 2003?

A. 10

B. 15

C. 20

D. 30

9. Which two years did the same number of boys and girls graduate?

A. 2003 and 2004

B. 2003 and 2007

C. 2004 and 2005

D. 2004 and 2006

Short Response

10. The data below shows the amount of money the band and the chorus raised during their candy drive over the past 4 months.

Months	Band	Chorus
January	$125	$170
February	$200	$190
March	$100	$100
April	$175	$75

Part A.

Based on the data in the table, create a double bar graph on the graph below. Be sure to

- title the graph
- label the axes
- graph all of the data
- provide a key

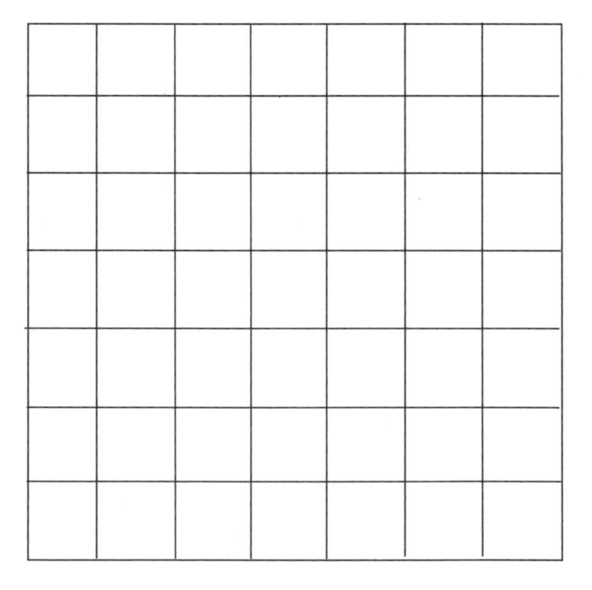

Part B.

Which group sold the most candy from January to April? Explain how you found your answer.

___

___

___

DOUBLE LINE GRAPHS

A **double line graph** is a line graph that uses two lines to make comparisons.

Example

The double line graph below compares the temperatures, in degrees F., of Chicago and Boston during the same week in March. On which day was the temperature difference between the two cities the greatest?

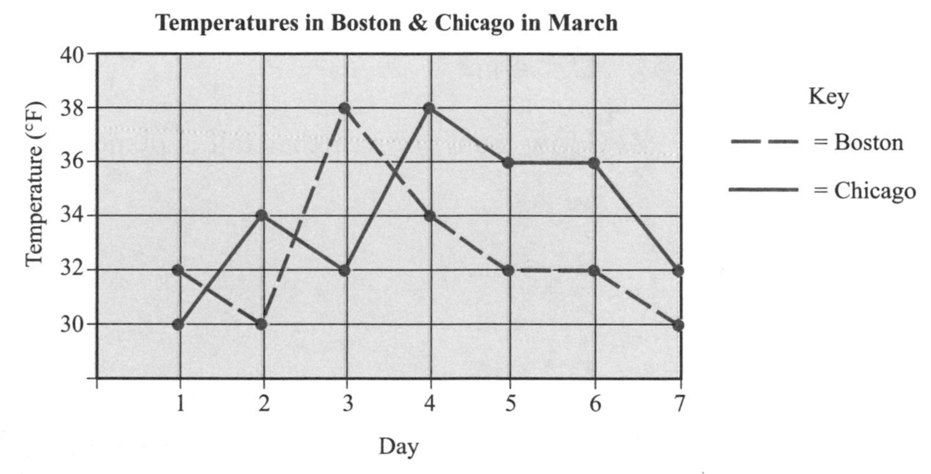

Solution:

To find the answer, find the longest gap between the two graphs.
The biggest gap is on Day 3.
Boston has the highest point on Day 3 at 38° F.
Chicago has the lowest point on Day 3 at 32° F.
There is a difference of 6° F on this day.
On Day 3 there was the biggest difference in temperature.

TEST YOUR SKILLS (For answers, see page 288.)

The double line graph below compares the height of two plants in inches. Use the double line graph to answer questions 1–3.

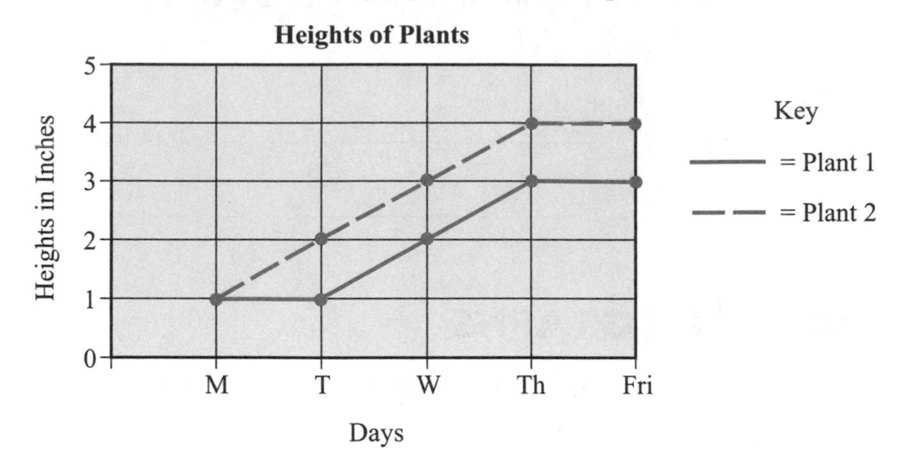

1. The double line graph shows the height of two plants in inches over the course of a week. How tall was plant 1 on Thursday?

   A. 1 in.

   B. 2 in.

   C. 3 in.

   D. 4 in.

2. How many more inches tall is plant 2 on Friday than plant 1?

   A. 1 in.

   B. 2 in.

   C. 3 in.

   D. 4 in.

3. What conclusion can you make based on the graph?

   A. Plant 1 is taller than plant 2.

   B. Plant 1 and plant 2 both started with a height of 3 inches.

   C. The height of both plants decreased during the week.

   D. The height of plant 2 increased 1 inch a day between Monday and Thursday.

The double line graph below shows CD and iTune sales from 2003 to 2007. Use the double line graph to answer questions 4–6.

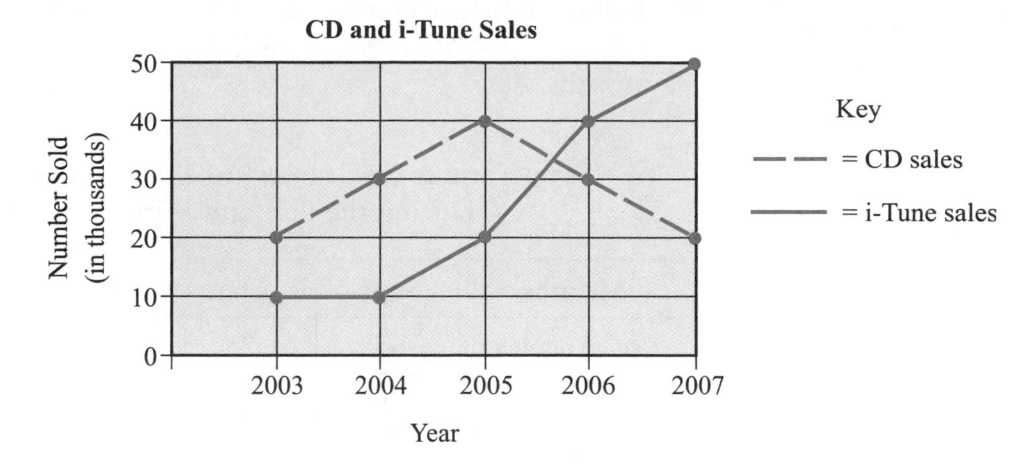

4. Between which two years did iTune sales increase the most?

   A. 2003 and 2004
   B. 2004 and 2005
   C. 2005 and 2006
   D. 2006 and 2007

5. How many more iTunes were sold in 2007 than CDs?

   A. 20,000
   B. 25,000
   C. 30,000
   D. 35,000

6. What conclusion can you make based on the graph?

   A. The number of CDs sold always increased.
   B. The number of iTunes sold increased from 2004 to 2007.
   C. The highest number of CDs sold was 50,000.
   D. The highest number of iTunes sold was in 2005.

Short Response

7. The table below compares the average high temperatures between Los Angeles and Buffalo during the fall and winter months.

Temperatures (in degrees) of Los Angeles and Buffalo
During the Fall and Winter Months

Months	Sept.	Oct.	Nov.	Dec.	Jan.
Los Angeles	65°	75°	75°	68°	65°
Buffalo	78°	65°	70°	45°	32°

Part A.

Draw a double line graph using the table above. Be sure to

- title the graph
- label the axes
- graph all of the data
- provide a key

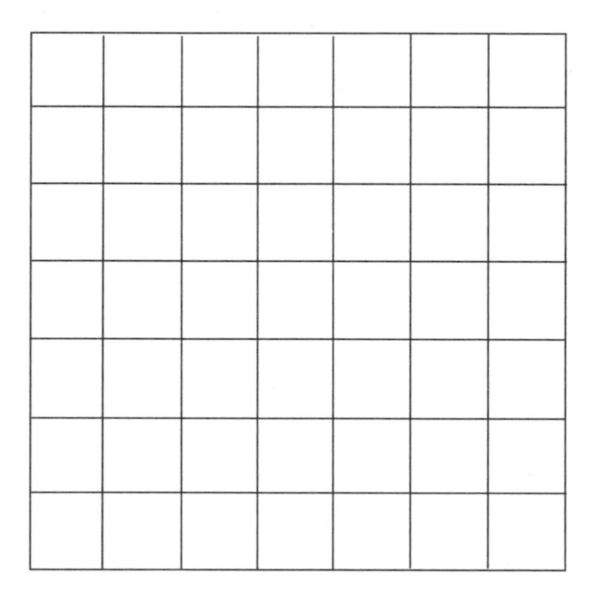

Part B.

Which month had the greatest temperature difference between the two cities? Show your work.

Answer __________

MISLEADING STATISTICS

In advertising, graphs are often used to persuade people's opinions. Based on how a graph is drawn, people may reach an incorrect conclusion about the data presented.

When reading graphs, it is important to make sure the graph is drawn correctly.

A graph is misleading if

- the y-axis (vertical axis) does not start at zero
- the y-axis (vertical axis) is not evenly spaced

It is also important to remember to make a conclusion based on the facts of the graph, and not based on your opinion.

Example

Look at the following graphs comparing the cost of gas for a Honda Civic (HC) and an SUV.

Gas Expense

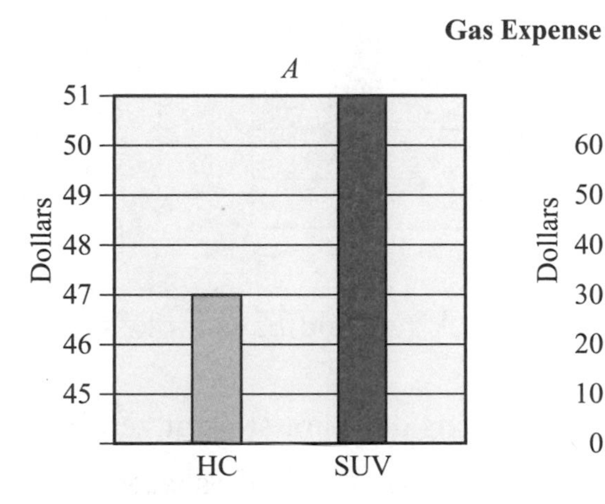

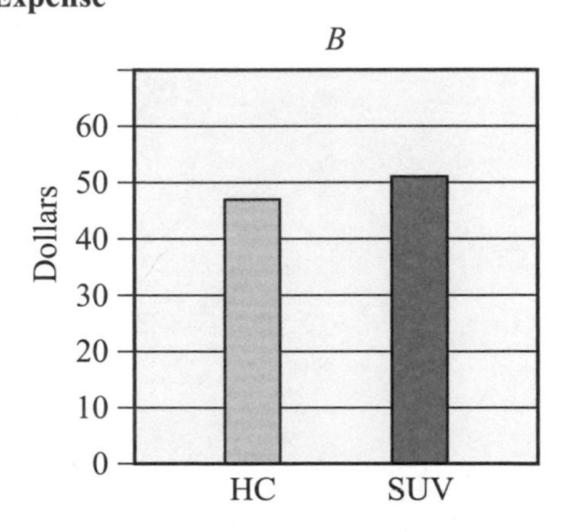

1. Which graph would you use to indicate that the cost of gas for an SUV is much greater than for a Honda Civic?

   Solution:

   The y-axis in Graph A does not start at zero, it starts at \$45. The y-axis in Graph B starts at zero. Both graphs show that gas for an SUV costs about \$4 more than gas for the Honda Civic, but Graph A visually makes it look like there is a larger difference between the costs of gas for both vehicles.

2. Why is Graph A misleading?

   A. The survey was taken when gas prices were high.

   B. The vertical axis is not evenly spaced.

   C. The vertical axis does not start at zero.

   D. It shows that all SUVs are more expensive to drive.

   Solution: C A graph is misleading if the y-axis does not start at zero.

TEST YOUR SKILLS (For answers, see page 289.)

1. Graphs A and B show the percent increase in sales for Papa Jack's Pizza in 2005 and 2006. Why is graph B misleading?

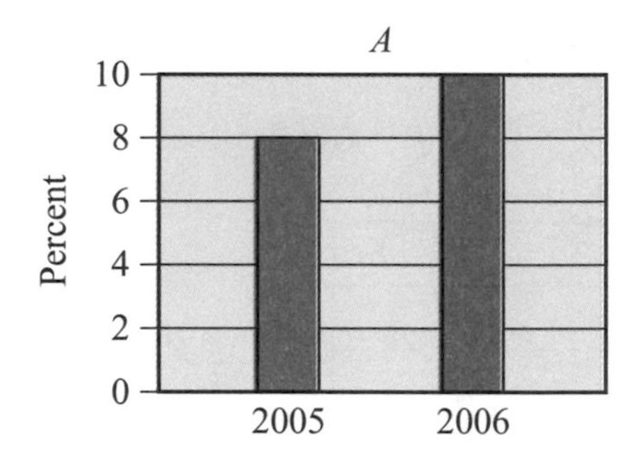

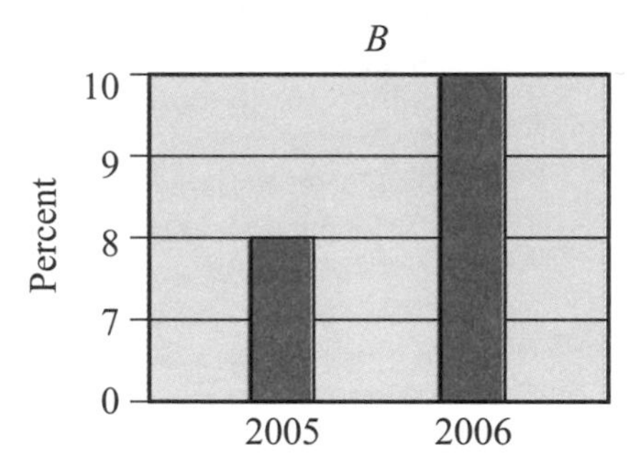

   A. Papa Jack's should have looked at more than two different years.

   B. The y-axis does not start at zero.

   C. The y-axis is not evenly spaced.

   D. The bars are not the same width.

2. A subway shop recorded the sizes of subs that people ordered for the day. Which conclusion is misleading?

Size	Number of Subs
Small	3
Medium	8
Large	15

A. Most people order large subs.

B. More people order medium subs than small subs.

C. People do not order small subs.

D. More people order large subs than small subs.

3. Why is the following graph misleading?

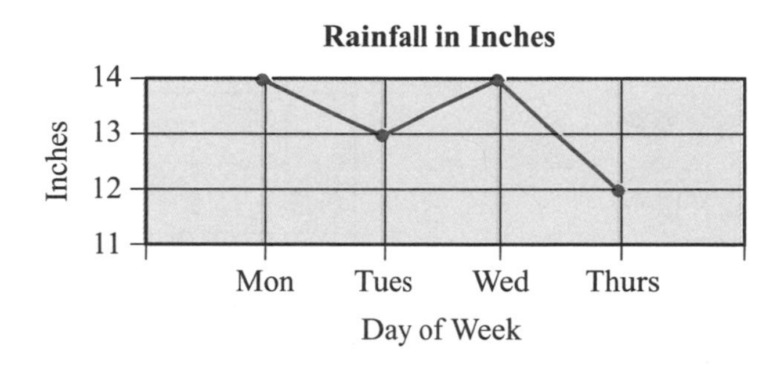

A. The days looked at should have been Friday, Saturday, and Sunday.

B. The graph is not titled.

C. The y-axis is not evenly spaced.

D. The y-axis does not start at zero.

4. Which conclusion could be drawn based on the following table?

FAVORITE FOOD

Hamburgers	6
Hot dogs	5
Nachos	11
Pizza	3

A. More people like pizza than hot dogs.

B. No one likes hamburgers.

C. Five more people like nachos than hamburgers.

D. Twice as many people like hot dogs than hamburgers.

Use the following graph to answer questions 5 and 6.

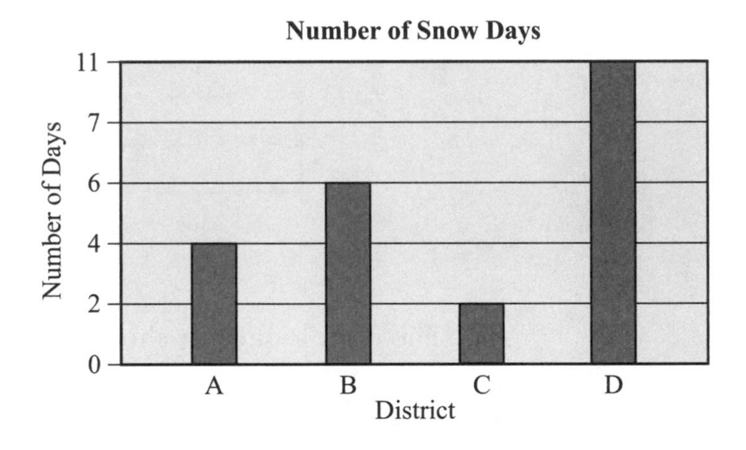

5. Why is the graph misleading?

A. The vertical scale uses intervals that are not equal.

B. The survey only interviewed sixth, seventh, and eighth graders.

C. The vertical scale does not start at zero.

D. The graph should be a line graph instead of a bar graph.

6. Which statement is a valid conclusion based on the graph?

   A. There was one more snow day in District A than District B.

   B. District C had the fewest number of snow days.

   C. District B had three snow days.

   D. There were twice as many snow days in District C than District B.

7. Which statement is valid based on the following graph?

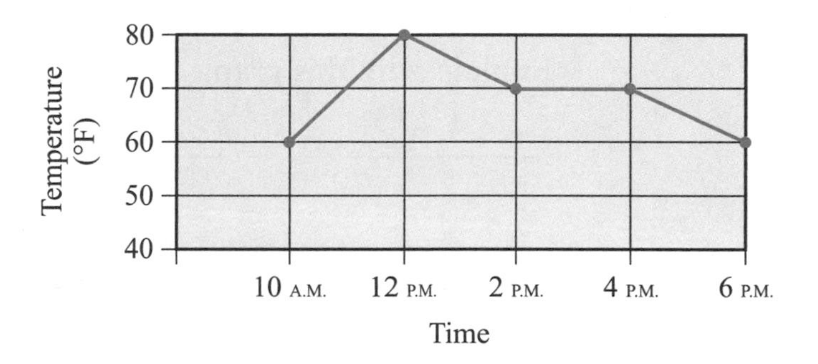

   A. The temperature was at its highest at 2:00 P.M.

   B. The temperature was above normal on this particular day.

   C. The difference in temperature from 10:00 A.M. to 12:00 P.M. was 20° F.

   D. At 6:00 P.M. the temperature was 50°.

Short Response

8. The bar graph below shows the average minutes Emily, Amber, Jenna, and Caitlin each spend talking on the phone.

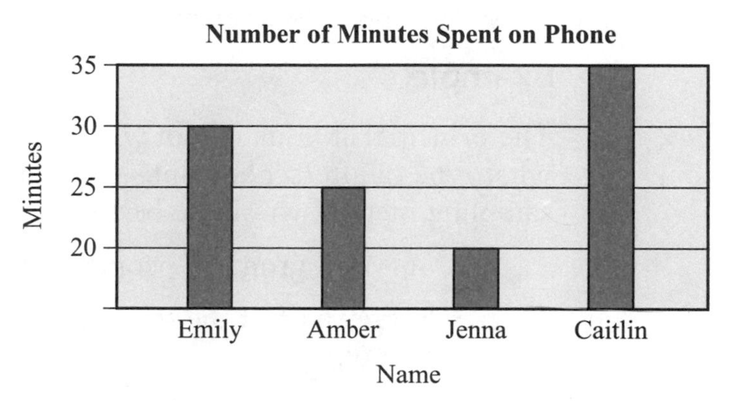

Part A.

Write a conclusion that you can draw from the graph about how many minutes Amber talked on the phone compared to Caitlin.

____

____

Part B.

Explain why this graph is misleading.

____

____

____

SAMPLING

Sampling is the process of collecting information from a part of a group. The sample should be representative of the larger group, known as the **population**.

To obtain a valid sample, the method of choosing the people should be **random** and **nonbiased**.

- Random—each person or object has an equally likely chance of being selected
- Nonbiased—does not favor one outcome or another

Example

The principal at your school wants to conduct a survey to determine what type of afterschool clubs the school should offer. Which sampling method would be best for the principal to use?

A. Survey a group of people at the mall.

B. Survey half of the students from each grade level in your school.

C. Call every fiftieth name in the phone book.

D. Survey the students at the local area college.

Solution:

For valid results, the principal should survey half the students from each grade level in your school (choice B). This sample best represents the type of students in the school. Each of the other choices do not target the population of your school, and therefore would not be helpful to your principal.

The above example shows how choosing a good sample is important when conducting a survey.

TEST YOUR SKILLS (For answers, see page 290.)

1. Brandon wants to conduct a survey to find out what students in his school list as their favorite movie. Who should he ask?

   A. every twentieth person at the local mall

   B. every fifteenth student walking into school

   C. the basketball players

   D. the kids in his neighborhood

2. Telecom wants to find out how many people per household own a cell phone in Wayne county. Which of these would be the best sampling method to use?

   A. survey the students at the local middle school

   B. survey people at the gas station

   C. survey people as they walk into the library

   D. call every tenth person in the Wayne county phone book

3. Cindy wants to find out how people feel about a neighborhood curfew. She decides to conduct a survey. Which method of sampling would give Cindy the best sample to use?

   A. ask students as they get off the school bus

   B. go to every other house in the neighborhood

   C. ask neighbors who are 85 years or older

   D. survey people as they walk into the local grocery store

4. Drew is writing an article for the newspaper club about people's favorite restaurants in his town. To complete his article, he stands outside Pauli's Pizza and asks people to name their favorite type of food. Which sampling technique would give Drew more accurate results for his article?

   A. surveying people at every pizza shop in town
   B. surveying every other person as they walk into the grocery store
   C. surveying people as they walk out of Hamburger Jack's
   D. surveying the students at his high school

5. Five hundred students attend Brown Middle School. A survey was given to 300 students to vote on the winter dance theme. After the surveys were collected, the first 50 votes were read and tallied. Based on the 50 votes, the theme for the dance is going to be Winter Wonderland. Why are the results of this survey invalid?

   A. only the girls in the school should have been surveyed
   B. only the boys in the school should have been surveyed
   C. all of the votes should have been tallied
   D. the first vote should have decided the theme of the dance

Short Response

6. Skyler is 12 years old. She wants to find out what type of music students her age listen to. She decides to conduct a survey of 11–14 year olds. After she makes a questionnaire, she distributes it to all of the homerooms in her school.

Part A.

Is this a good sampling technique to use?

Answer __________

Part B.

Explain why or why not.

____

____

____

Chapter 7

PRACTICE TEST 1

MULTIPLE CHOICE (Circle the correct answer.)

1. What is the solution of the equation $2x - 3 = 17$?

   A. $x = 7$
   B. $x = 8$
   C. $x = 9$
   D. $x = 10$

2. Write 502,000 in scientific notation.

   A. 5.02×10^4
   B. 5.02×10^5
   C. 50.2×10^4
   D 502×10^3

3. Emily sold chocolate chip, peanut butter, sugar, and oatmeal cookies. She wants to make a graph of her data. Which type of graph should Emily use to best represent the different types of cookies that she sold?

   A. circle graph
   B. bar graph
   C. line graph
   D. histogram

4. John is running a 5-kilometer race. How many meters long is this?

A. .005

B. 50

C. 500

D. 5000

5. What is the greatest common factor of 6, 18, and 36?

A. 2

B. 3

C. 6

D. 9

6. Simplify the expression: $2^3 \cdot 2^2$

A. 2^5

B. 2^6

C. 2^7

D. 2^8

7. These are the number of items sold at a football concession stand during seven games. What is the range of the data?

16, 32, 108, 47, 112, 136, 82

A. 66

B. 98

C. 120

D. 168

8. Evaluate the expression if $w = 2$, $x = 5$, and $y = 3$.

$$xy - wx$$

A. 5

B. 9

C. 25

D. 65

9. Solve: $7x \geq 91$

A. $x \leq 13$
B. $x \geq 13$
C. $x \geq 84$
D. $x \geq 637$

10. What is the area of the figure on the grid below?

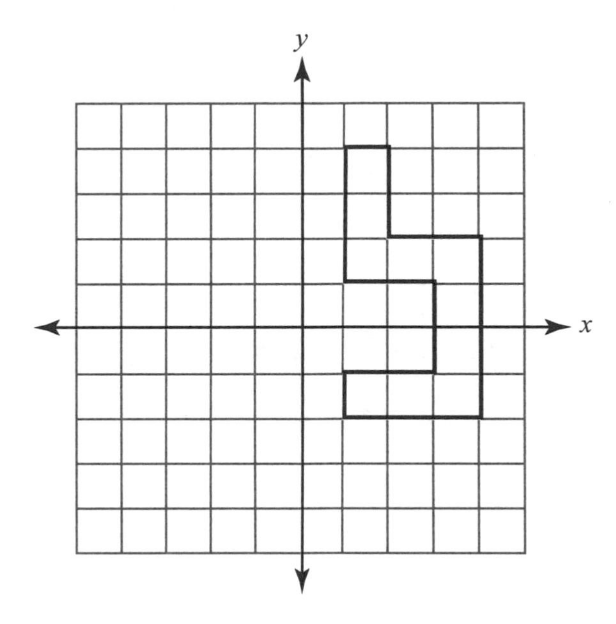

A. 6
B. 10
C. 13
D. 18

11. Find 7 – (–15).

A. –6
B. 6
C. –22
D. 22

12. What is the volume of a rectangular prism with a length of 14 inches, a width of 10 inches, and a height of 6 inches?

    A. 30 cu. in.

    B. 144 cu. in.

    C. 840 cu. in.

    D. 7000 cu. in.

13. Write the phrase "two less than four times a number" as an algebraic expression.

    A. $2 - 4n$

    B. $2 + 4n$

    C. 4 + n –2

    D. $4n - 2$

14. Which number is the greatest?

    A. 1.045×10^4

    B. 2.52×10^4

    C. 3.15×10^3

    D. 7.2×10^3

15. Reed has a square sandbox in his backyard that has an area of 18 square feet. Estimate the length of the side of the sandbox to the nearest whole number.

    A. 4 ft.

    B. 5 ft.

    C. 9 ft.

    D. 20 ft.

16. To which set of numbers does $\sqrt{25}$ *not* belong?

    A. natural

    B. integers

    C. rational

    D. irrational

17. Find the missing angle.

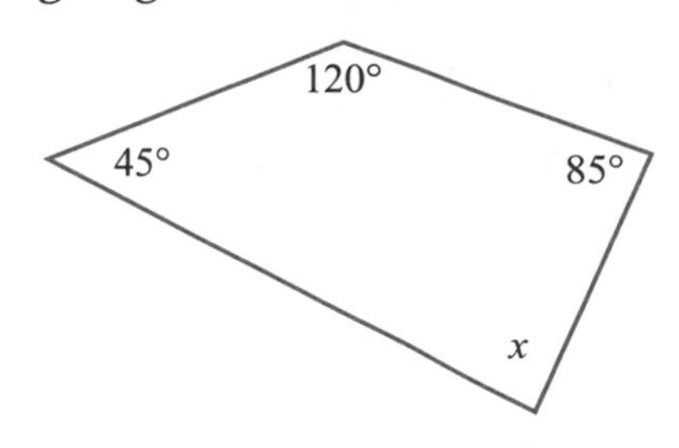

A. 45°

B. 85°

C. 110°

D. 120°

18. In a survey, students were asked to name their favorite school subject. The results are shown in the table below.

Favorite Subject

History	5
Math	8
Science	3
ELA	6

Which conclusion can be drawn from the table?

A. More students like history than math.

B. Half of the students like math.

C. More students like math because they like the teacher.

D. Half of the students like math and science.

19. Which measure of central tendency best describes the data given below?

4, 8, 12, 16, 35, 87

A. mean

B. median

C. mode

D. range

20. Find the least common multiple of 4, 6, and 18.

A. 2
B. 24
C. 36
D. 108

21. The $\sqrt{79}$ lies between which two consecutive integers?

A. 7 and 8
B. 8 and 9
C. 9 and 10
D. 10 and 11

22. Which point on the graph would make the figure a rectangle?

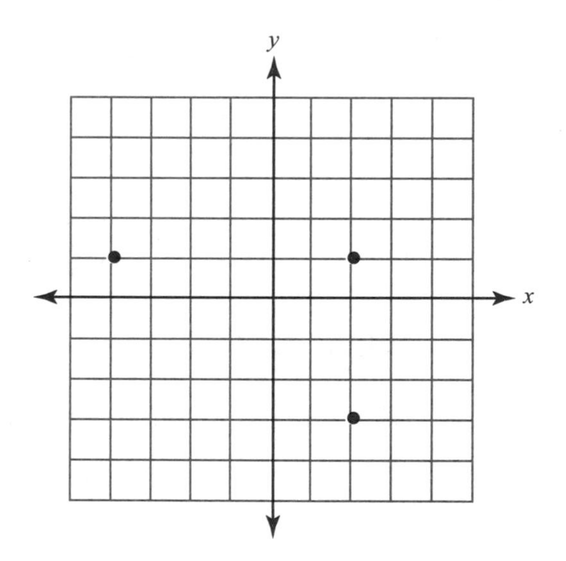

A. (−3, −4)
B. (−4, −3)
C. (−4, 3)
D. (4, 3)

23. Which metric unit of measure would you use to measure the distance across a state?

A. inches

B. kilometers

C. yards

D. miles

24. What is the name of the figure shown below?

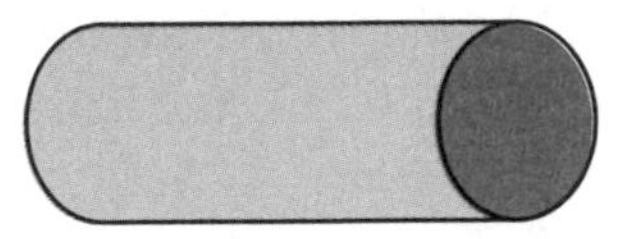

A. cone

B. cylinder

C. rectangular prism

D. pyramid

25. How many liters are in 5 kiloliters?

1 kiloliter = 1000 liters

A. .05

B. 5

C. 500

D. 5000

26. Which of the following is the most reasonable estimate of $36.04 \div 5.96$?

A. 5

B. 6

C. 7

D. 8

27. For every 10 fluid ounces of laundry detergent, you can do six loads of laundry. How many loads of laundry can you do if you have 25 fluid ounces of detergent?

A. 5

B. 12

C. 15

D. 21

28. How many outcomes are there for spinning a spinner labeled A, B, C, D and tossing a quarter?

A. 6

B. 8

C. 12

D. 36

29. Which scale should be used to measure the mass of a bag of grapes?

A. balance

B. bathroom scale

C. truck scale

D. produce scale

30. A garage door keypad has 9 digits. The code to open or close the door has to be 3 digits. How many possible code combinations are there?

A. 24

B. 27

C. 500

D. 729

SHORT RESPONSE

31. Part A.

Solve the inequality. Show your work.

$$\frac{x}{4} \leq -2$$

Answer __________

Part B.

Graph the solution on the number line below.

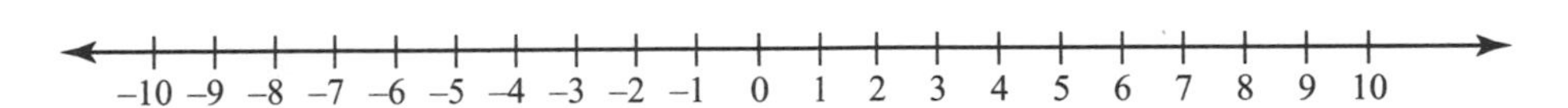

32. In Ms. Tiffin's math class, students are learning about prime factorization. The warm-up problem is to find the prime factorization of 504.

Part A.

Find the prime factorization of 504. Show your work.

Answer __________

Part B.

Write your answer to Part A in exponential form.

Answer __________

33. Kelly is 2 years younger than three times Tracy's age.

Part A.

If a represents Tracy's age, write an algebraic expression that can be used to find Kelly's age.

Answer __________

Part B.

If Tracy is 12 years old, how old is Kelly? Show your work.

Answer __________

Part C.

If Kelly is 16 years old, how old is Tracy? Show your work.

Answer __________

34. Mike wants to wrap this shoebox with holiday paper. Estimate the amount of paper Mike will need in square inches. Show your work.

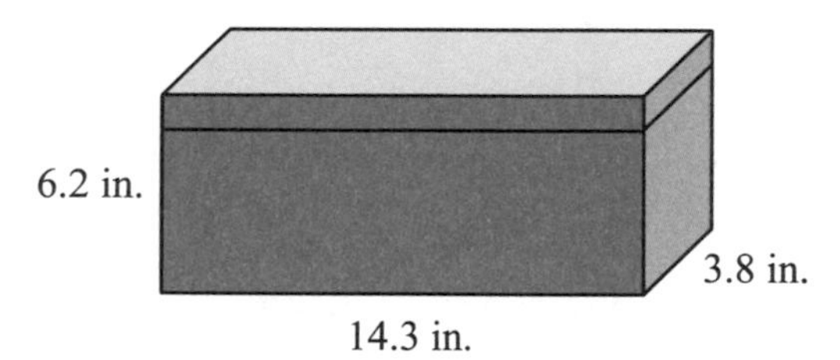

Answer __________

35. The seventh graders at Armstrong Middle School went to Flags Amusement Park for a Math and Science Field Day. They know that the distance around the merry-go-round is 27 feet. Find the diameter of the merry-go-round to the nearest tenth. **Show your work.**

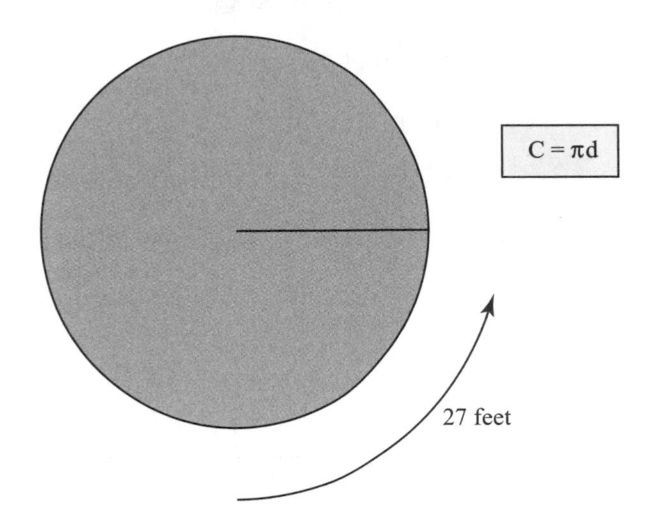

Answer __________

36. Breann's parents have decided to record the number of text messages that she sends to her friends during the week. The table below shows the results over a two-week period.

Number of Text Messages

Day of the Week	Week 1	Week 2
Monday	22	17
Tuesday	36	23
Wednesday	18	6
Thursday	25	32
Friday	41	39

Part A.

Make a double line graph using the given data.

Be sure to

- title the graph.
- label the axes.
- graph all of the data.
- provide an appropriate key.

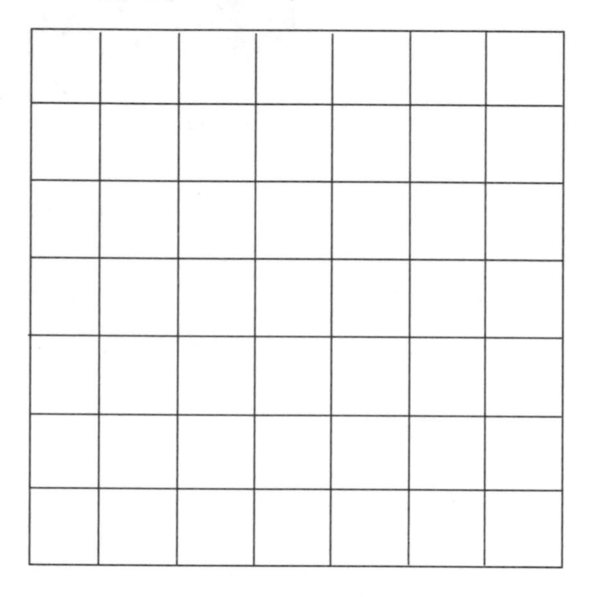

Part B.

Write one conclusion that you can make based on the graph.

___

___

Part C.

What is the difference between the average number of text messages Breann sent during Week 1 and the average number of text messages sent during Week 2? Show your work.

Answer ________

37. Tommy just bought a small bag of Skittles. The table below shows the number of Skittles of each color in the bag.

Skittle Color	Number of Skittles
Red	5
Orange	8
Yellow	2
Green	6
Purple	4

Part A.

Tommy randomly chooses a Skittle and does not return it to the bag. He then chooses another Skittle. What is the probability that Tommy chooses two yellow Skittles?

Show your work.

Answer __________

Part B.

On the lines below, explain why your answer in Part A is correct.

____

____

____

____

38. Madison wants to repaint her bedroom. She knows that one gallon of paint will cover about 350 square feet of wall space.

Part A.

Madison measured her walls and found that she has 2450 square feet of wall space in her bedroom. How many gallons of paint will Madison need? Show your work.

Answer __________

Part B.

At the paint store, Madison is told that they only sell the color that she wants in quarts! How many quarts of paint will she need? Show your work.

Answer __________

Chapter 8

PRACTICE TEST 2

MULTIPLE CHOICE (Circle the correct answer.)

1. What is the prime factorization of 90?

A. $2 \times 3 \times 5$
B. $2 \times 3^2 \times 5$
C. $2^2 \times 3 \times 5$
D. $2 \times 5 \times 9$

2. How many pints are in 2 gallons?

4 quarts = 1 gallon 2 pints = 1 quart

A. 2 pts.
B. 8 pts.
C. 16 pts.
D. 20 pts.

3. What are the first four factors of 56?

A. 1, 2, 4, 7
B. 1, 2, 4, 8
C. 2, 3, 4, 7
D. 2, 4, 6, 7

4. How many faces does this solid figure have?

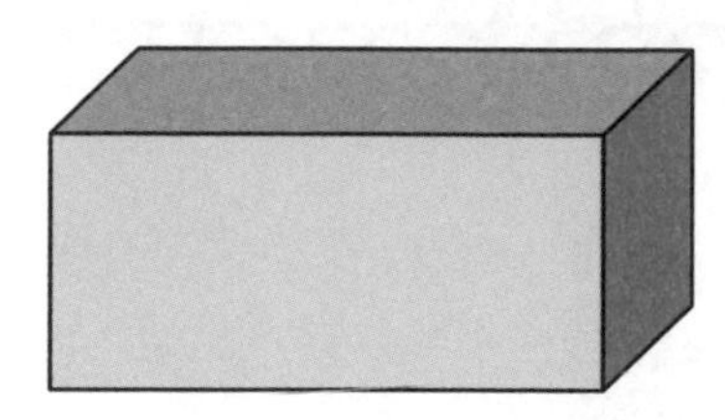

A. 2
B. 4
C. 6
D. 8

5. Which of the following is a whole number?

A. $-\sqrt{25}$
B. 3
C. .25
D. $\frac{1}{2}$

6. Write 2.4×10^3 in standard form.

A. .0024
B. 240
C. 2400
D. 24,000

7. Simplify: $|-20| + 15 \div 5 \cdot 2^2$

A. 28
B. 32
C. 56
D. 92

8. Solve: $\frac{x}{4} - 3 = 8$

A. 5

B. 35

C. 44

D. 96

9. Evaluate: $2^6 \div 2^3$

A. 2

B. 8

C. 16

D. 32

10. Jack has a bag of 5 marbles; 2 red, 1 blue, and 2 yellow. He chooses a marble and then replaces it. He then picks another marble. What is the probability that Jack will choose 2 red marbles?

A. $\frac{4}{25}$

B. $\frac{4}{5}$

C. $\frac{2}{5}$

D. $\frac{4}{10}$

11. You are playing flag football with your friends. During three plays in the game you ran 10 yards, lost 15 yards, and then ran 25 yards. How many yards did you run at the end of the three plays?

A. 20 yds.

B. 25 yds.

C. 30 yds.

D. 40 yds.

12. Two medium pizzas from Pizza Paul's costs $9. How much will it cost if you want to order five medium pizzas?

A. $4.50

B. $7.50

C. $18.00

D. $22.50

13. At summer camp, everyone is assigned breakfast, lunch, and dinner clean-up duty. Based on the Venn diagram below, which of the following statements is true?

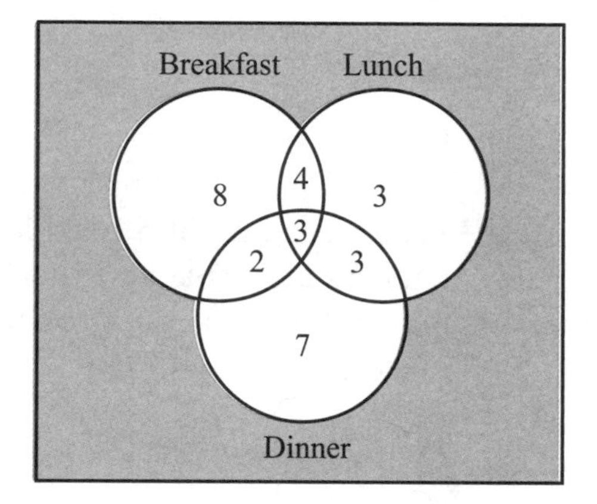

A. Eight students have dinner duty only.

B. Five students have breakfast, lunch, and dinner duty.

C. Twenty students have lunch duty.

D. Two students have breakfast and dinner duty.

14. What is the relative error of a pipe that is 12 inches long ± 0.24 inches?

A. 0.2

B. 0.02

C. 11.76

D. 12.24

15. The sum of three times x and eleven is equal to fourteen is represented by which equation?

A. $3 + 11x = 14$

B. $3 + 11 + x = 14$

C. $3(x + 11) = 14$

D. $3x + 11 = 14$

16. The table below shows the number of students signed up for afterschool activities.

Activity	Number of Students
Newspaper Club	15
Intramurals	32
Skateboard Club	13

What is the experimental probability that a student is signed up for the newspaper club?

A. $\frac{13}{60}$

B. $\frac{8}{15}$

C. $\frac{1}{4}$

D. $\frac{47}{60}$

17. Which number is a natural and whole number?

A. 3

B. -1

C. 0

D. $\frac{1}{2}$

18. Solve: $10 > x - 8$

A. $x < 2$

B. $x > 2$

C. $x < 18$

D. $x > 18$

19. A can of tomato soup has a diameter of 8 centimeters and a height of 14 centimeters. Find the volume of the can. Leave your answer in terms of π.

A. 56π cm^3

B. 112π cm^3

C. 224π cm^3

D. 896π cm^3

20. Name the central angle.

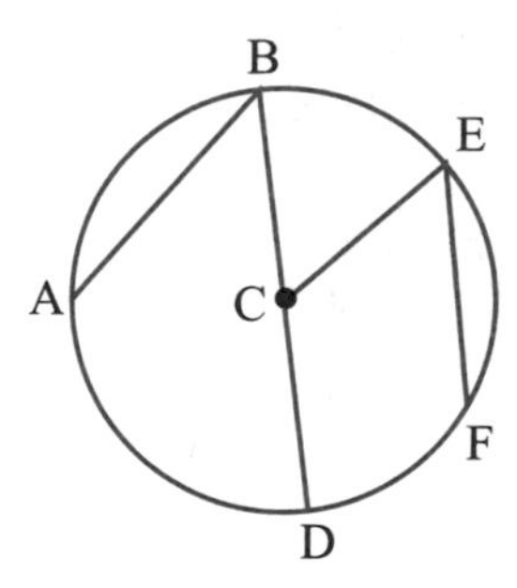

A. $\angle ABD$

B. $\angle CEF$

C. $\angle BCE$

D. $\angle CED$

21. Damon has a spinner that is divided into five equal parts. If he spins the spinner 160 times, what is a good prediction of how many times he can expect to get an even number?

A. 10

B. 32

C. 64

D. 100

22. Express in decimal form: 10^{-3}

A. 0.001

B. 0.01

C. 0.1

D. 10

23. Every seventh person through the gate at Dwyer Stadium receives a baseball and every twelfth person receives a hat. Which customer will receive a baseball and a hat?

A. 36

B. 84

C. 98

D. 108

24. The Knights scored the following points during their season. What is the range of their scores?

56, 43, 21, 96, 76, 98, 47

A. 21

B. 55

C. 75

D. 77

25. The bar graph below shows the number of points Katherine scored in four basketball games.

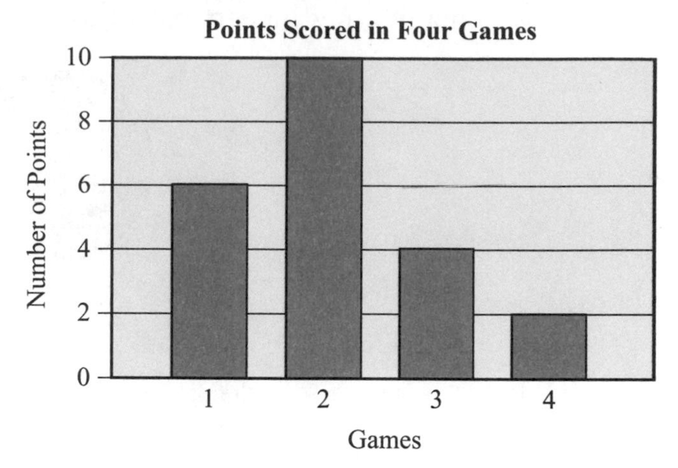

What conclusion can you make based on the graph?

A. Katherine scored twice as many points in Game 2 than Game 1.

B. Katherine scored the most points in Game 4.

C. Katherine scored eight more points in Game 2 than Game 3.

D. Katherine scored three times more points in Game 1 than Game 4.

26. In a survey, students were asked to name their favorite day of the week. The results are shown in the table below.

Day	Tally
Monday	|||
Tuesday	𝍸
Wednesday	||
Thursday	|
Friday	𝍸 ||||

What percent of the students surveyed said Friday was their favorite day of the week?

A. 25%

B. 35%

C. 45%

D. 60%

27. The formula for the volume of a rectangular prism is $V = lwh$. What is the height of the rectangular prism if the volume is 108 cubic inches, the length is 9 inches, and the width is 3 inches?

A. 3 in.

B. 4 in.

C. 12 in.

D. 2916 in.

28. The circumference of a circle is 43.96 inches. Find the diameter of the circle. Use 3.14 for π.

A. 7 in.

B. 8 in.

C. 14 in.

D. 16 in.

29. Evaluate the expression when $m = 6$, $n = 2$, and $s = 3$.

$$4m + 3n - 5s + 3$$

A. 11

B. 18

C. 24

D. 54

30. Sean wants to conduct a survey to find out how many students in his middle school watch football on Sundays. Which of these would be the best method to use to obtain his sample, if his middle school includes grades 6–8?

A. Have sixth-, seventh-, and eighth-grade students complete surveys during their lunch periods.

B. Survey students who participate in sports.

C. Survey the students on the football team.

D. Survey the sixth-grade students.

SHORT RESPONSE

31. Order the following numbers from least to greatest.

4.3×10^2 3^3 10.5 $\frac{13}{2}$ $\sqrt{7}$ $\sqrt{121}$ 4300

Part A.

Show your work.

Answer ________

Part B.

Is $\sqrt{121}$ a rational or irrational number? Explain your answer on the lines below.

____

____

____

32. The Talking Network charges $25 a month for phone service plus $0.15 a minute for long distance calls.

Part A.

Let m represent the number of minutes Rose talked on the phone long distance. Write an equation you could use to solve for the number of minutes Rose was on the phone long distance if her phone bill for the month was $100.

Answer ________

Part B.

Solve your equation to find out how many minutes Rose was on the phone long distance. Show your work.

Answer ________

33. Rhonda surveyed a group of her friends to find out which season is their favorite season. Her results are shown in the table below.

SEASON	NUMBER OF FRIENDS
Spring	4
Summer	12
Fall	6
Winter	3

Part A.

Using your protractor, make a circle graph of the given data.

Be sure to include:

- All of the data given
- Labeled sections for each part of the circle
- A title for your graph

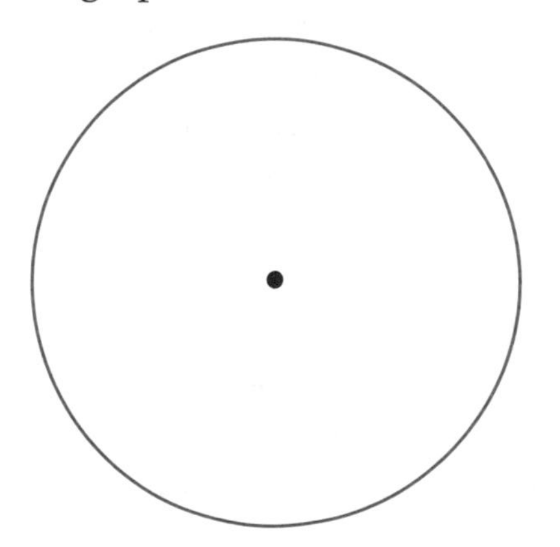

Part B.

What percent of the people surveyed said that summer was their favorite season?

Answer __________

34. The area of a circular picture frame is 28.26 square inches. Find the diameter of the frame to the nearest tenth. Use 3.14 for π. Show your work.

Answer ________

35. The formula for converting temperature from degrees Celsius to degrees Fahrenheit is $F = \frac{9}{5}C + 32$. What is the temperature in degrees Fahrenheit when the temperature is 25° Celsius?

Part A.

Show your work.

Answer ________

Part B.

The temperature in Toronto, Canada, on July 11th was 20° Celsius. It was 82° Fahrenheit in Rochester, New York, on the same day. What was the temperature difference between the two cities? Show your work.

Answer ________

36. Find the surface area of a cylinder with a radius of 6 millimeters and a height of 12 millimeters. Use 3.14 for π. Show your work.

Answer ________

37. The Pierces are going to put a new fence around their square-shaped yard. The area of the yard is 1024 square feet.

Area = $1024\ ft^2$

Part A.

How much fencing, in feet, do the Pierces need to fence around their square yard? Show your work.

Answer __________

Part B.

Fencing is sold in lengths of 10 feet. Estimate how many lengths of fencing the Pierces will need to buy to fence in their yard. Show your work.

Answer __________

38. During the first five weeks of school, Scott earned scores of 80, 100, 60, 100, and 70 on his math tests.

Part A.

Find the mean, median, mode, and range of Scott's test scores. Show your work.

Mean __________ Median __________

Mode __________ Range __________

Part B.

If Scott wants to impress his parents, which measure of central tendency should he use?

____

____

____

Part C.

Explain why the measure of central tendency used in Part B might be considered misleading. Which would be a more accurate measure to use?

____

____

____

Chapter 9

TEST YOUR SKILLS SOLUTIONS

CHAPTER 1: NUMBER SENSE AND OPERATIONS

OPERATIONS WITH INTEGERS

1. B $-25 + -36 = -61$
2. D $-23 + 7 = -16$
3. A $-6 + 15 = 9$
4. C $48 - (-24) = 48 + 24 = 72$
5. C $-8(9) = -72$
6. A $-4(-11) = 44$
7. C $(-2)(3)(-4) = (-6)(-4) = 24$
8. B $\frac{-36}{-4} = 9$
9. A $-72 \div 9 = -8$
10. D Start at 6° and keep subtracting 3 until you get to −12 (3, 0, −3, −6, −9, −12). It will take 6 hours.
11. D $\frac{-12}{-3} = -4$ is not a true statement.

Short Response

12. Part A. −14

$-8 + 3 - 10 - 4 + 5$	Write the integers that represent the points for each day.
$-8 + 3 + (-10) + (-4) + 5$	Change all of the subtraction to addition of the opposite.

$-22 + 8$ — Combine all negatives together and all positives together.

-14 — Simplify.

Part B. 14 — Since the stock is at -14, it will take 14 points to reach zero.

EXPONENTS

1. **B** Since multiplying, add the exponents or
   $2^3 \cdot 2^4 = 2 \cdot 2 \cdot 2 \cdot 2 \cdot 2 \cdot 2 \cdot 2 = 2^7$
2. **D** $3^2 \cdot 3^3 = 9 \cdot 27 = 243$
3. **B** Since this is a power to a power, multiply the exponents
   $(2^3)^2 = 2^6$
   or write out as a multiplication problem and add the exponents.
   $(2^3)^2 = 2^3 \cdot 2^3 = 2^6$
4. **C** When dividing, since bases are the same, subtract the exponents.
   $$\frac{4^3}{4^1} = 4^2 = 16$$
5. **C** $\frac{5^4}{5^2} = 5^2 = 25$
6. **B** Since it is a power to a power, multiply the exponents.
   $(4^5)^2 = 4^{10}$
7. **A** $3^2 = 9$ and $2^3 = 8$. $9 > 8$

Short Response

8. **Part A.**

Start with one cell.	0	2^0
At 10 minutes there will be two cells.	0 0	2^1
At 20 minutes each of these cells splits into two more cells.	0 0 0 0	2^2
At 30 minutes, the four cells each split into two more cells.	0 0 0 0 0 0 0 0	2^3

There will be eight cells in 30 minutes.

Part B.

The cell will split into two cells every 10 minutes. There are three sets of 10 minutes in 30 minutes. So $2^3 = 8$.

ORDER OF OPERATIONS WITH ABSOLUTE VALUE, INTEGERS, AND EXPONENTS

1. B $|-5| + 9 \times 6 \div 2$ Take the absolute value of 5.
$5 + 9 \times 6 \div 2$ Multiply.
$5 + 54 \div 2$ Divide.
$5 + 27$ Add.
32

2. A $4(8 - 3) + 3(2)^2$ Do the subtraction inside the parentheses.
$4(5) + 3(2)^2$ Simplify the exponent.
$4(5) + 3(4)$ Multiply left to right.
$20 + 12$ Add.
32

3. B $\frac{(10-6)^2}{2(4)}$ Do the subtraction inside the parentheses.
$\frac{4^2}{2(4)}$ Simplify the exponent.
$\frac{16}{2(4)}$ Multiply in the denominator.
$\frac{16}{8}$ Divide.
2

4. D $6(9 - 3) \div 3(4)$ Do the subtraction inside the parentheses.
$6(6) \div 3(4)$ Multiply (need to multiply and divide left to right).
$36 \div 3(4)$ Divide (need to go left to right!).
$12(4)$ Multiply.
48

5. B $3(2^3 - 2^2) + 3(4)$ Simplify the exponents in the parentheses.
$3(8 - 4) + 3(4)$ Subtract in the parentheses.
$3(4) + 3(4)$ Multiply left to right.
$12 + 3(4)$ Multiply.
$12 + 12$ Add.
24

6. **A** Addition is the first operation performed inside the parentheses.

7. **C** $\frac{4(15-3)}{2(3)}+5$

$\frac{4(12)}{2(3)}+5$

$\frac{48}{6}+5$

$8+5$

13

Short Response

8. **Part A.** Multiplication.

The operation that is performed first is the multiplication inside the parentheses (because of the exponent). Note—exponent is *not* an operation.

Part B. 36

$2(3^3-3^2)$

$2(27-9)$

$2\ (18)$

36

NEGATIVE AND ZERO EXPONENTS

1. **D** $10^3 = 10 \cdot 10 \cdot 10 = 100 \cdot 10 = 1000$

2. **C** $10^{-2} = \frac{1}{10^2} = \frac{1}{100}$

3. **B** Any number raised to the zero power equals 1. $5^0 = 1$

4. **C** $(-10)^4 = (-10)(-10)(-10)(-10) = (100)(-10)(-10) = (-1000)(-10) = 10{,}000$

5. **B** $10^{-3} = \frac{1}{10^3} = \frac{1}{1000} = .001$

6. **C** $2(3^0) = 2(1) = 2$

Short Response

7. Change all of the numbers to decimals and then order them.

10^{-2}	100	.0001	10	$\frac{1}{10}$
$\frac{1}{100}$				
0.01	100	.0001	10	0.10
#2	#5	#1	#4	#3

Answer: .0001, 10^{-2}, $\frac{1}{10}$, 10, 100

SQUARE ROOTS AND ESTIMATING SQUARE ROOTS

1. A $\sqrt{36} = 6$
2. C $\sqrt{25} = 5$, $\sqrt{36} = 6$, $\sqrt{64} = 8$. The $\sqrt{55}$ is *not* a perfect square.
3. B $\sqrt{121} = 11$
4. A The area of the square is 144 square units. Since the sides of a square are equal and the area is side times side, you have to find a number that when multiplied by itself equals 144. An easier solution is to take the square root of 144.

   12 [Area = 144] 12

   $A = s^2$
   $144 = s^2$
   $\sqrt{144} = s$
   $12 = s$

5. C $\sqrt{64} = 8$ and $\sqrt{81} = 9$, so the $\sqrt{80}$ is between 8 and 9.
6. B $\sqrt{121} = 11$ and $\sqrt{144} = 12$, so the $\sqrt{123}$ is between 11 and 12.

Short Response

7. **Part A.** 4, 9, 25, 121

   Part B. 4, 9, 25, and 121 are perfect squares because each of the numbers can be multiplied by a number and itself. Examples: 2(2) = 4, 3(3) = 9, 5(5) = 25 and 11(11) = 121.

   Part C. There are many possible solutions. Some examples are 51, 52, 53, 54, 60.

SUBSETS OF NUMBERS

1. **A** The set of whole numbers is {0, 1, 2, 3, 4, ...}
2. **B** An irrational number is a number that does not repeat and does not end.
3. **C** The set of whole numbers are the numbers 0, 1, 2, 3, and so on.
4. **C** An integer is a positive or negative whole number. { ...–3, –2, –1, 0, 1, 2, 3, ...} –5 is included in this set.
5. **A** A rational number is a number that can be written as a fraction, or is a decimal that terminates. 0.65 is the only choice that is a decimal that ends.
6. **A** The set of natural numbers is {1, 2, 3, 4, ...}. Zero is not included in this set.
7. **D** The $\sqrt{5}$ is a decimal that does not end and does not repeat. This makes it an irrational number.
8. **B** $-1\frac{2}{5}$ is a rational number because it is a number that can be written as a fraction.

Short Response

9. **Part A.**

RATIONAL	IRRATIONAL
-3	π
$\sqrt{16}$	$\sqrt{13}$
$\frac{1}{2}$	$-\sqrt{15}$

Part B.

The number 3.122333444455555 ... is irrational because it is a number that does not repeat the same digits (for example: 3.12121212 ...) and it goes on forever.

RATIONAL AND IRRATIONAL NUMBERS ON A NUMBER LINE

1. **D** $\sqrt{49} = 7$. Letter *M* shows a point on the number 7.

2. **B** $\frac{1}{2} = 0.5$. There should be a point between 0 and 1, and there is not.

3. **C** $-\pi = -3.14$. -3.14 is between -3 and -4. Graph C shows the point graphed correctly.

4. **A** $\sqrt{49} = 7$, so the $\sqrt{50}$ is a little bigger than 7. Graph A shows the point graphed correctly.

5. **C** The point is between 3 and 4. Since $\sqrt{9} = 3$ and $\sqrt{16} = 4$, the $\sqrt{12}$ would be between 3 and 4.

6. **D** From least to greatest: -0.5, $\sqrt{5} = 2.2\ldots$, $\frac{8}{2} = 4$

7. **B** The points on the graph, from greatest to least are *K*, *J*, *I*, *H*. The numbers that match up with these letters would be $K = 5.75$, $J = \sqrt{16} = 4$, $I = \sqrt{3}$, which is between 1 and 2, and $H = -\frac{1}{4} = -0.25$.

Short Response

8. **Part A.** Change each number to a decimal:

$-\frac{1}{2} = -0.5$ $\quad \sqrt{9} = 3$ $\quad -\sqrt{4} = -2$ $\quad \pi = 3.14\ldots$ $\quad \frac{1}{4} = 0.25$

On the number line:

$-\sqrt{4}$ $\quad -1/2$ $\quad 1/4$ $\quad \sqrt{9}$ $\quad \pi$

-3 $\quad -2$ $\quad -1$ $\quad 0$ $\quad 1$ $\quad 2$ $\quad 3$ $\quad 4$ $\quad 5$

Part B. $\pi, \sqrt{9}, \frac{1}{4}, -\frac{1}{2}, -\sqrt{4}$

REPEATING AND NONREPEATING DECIMALS

1. **D** Change each fraction to a decimal. $\frac{1}{9} = 0.11111...$
2. **A** 0.75 is terminating because it is a number that ends; it does not go on forever. It is rational because you could write 0.75 as a fraction: $\frac{75}{100} = \frac{3}{4}$
3. **C** Change each number to a decimal. $\frac{4}{5}$ is the only decimal that ends since $\frac{4}{5} = 0.8$.
4. **A** Change each number to a decimal. $\sqrt{3}$ is the only decimal that does not end and therefore is nonterminating.
5. **A** Change each number to a decimal. 3π is the only number that does not repeat or end since $-3\pi = -9.4247779...$
6. **C** $\frac{9}{11}$ is a fraction so it is a rational number. If you convert it to a decimal (0.818181...) it repeats.
7. **A** $\sqrt{49} = 7$. The number 7 is a rational and terminating number because you can write 7 as a fraction, $\frac{7}{1}$, and it stops; it does not go on forever.

Short Response

8. **Part A.** $\frac{2}{5} = 0.4$ as a decimal. It is a terminating decimal because it stops.

   Part B. $\frac{1}{6} = 0.16666...$ as a decimal. It is nonterminating because it is a repeating decimal that does not end.

JUSTIFY REASONABLENESS USING ESTIMATION

1. **B** Round each number to the nearest mile. 23 + 29 + 33 = 85 miles.

2. **D** Round \$7.99 to \$8. Since there are four T-shirts, $4(8) = \$32$.

3. **D** Round \$3.23 to \$3.00. Since it is a 15-gallon tank, $15(3) = \$45$. Since this is an estimate, \$48 is the best choice.

4. **B** Round each amount to the nearest dollar. Your lunch = \$2 + \$1 = \$3. Kyle's lunch = \$3 + \$2 = \$5. Since your lunch was \$3, and Kyle's was \$5, Kyle spent \$2 more on lunch.

5. **A** Round each number to the nearest whole number, then divide. $\frac{112}{4} = 28$

6. **B** 296 pieces of candy ≈ 300 and 27 students ≈ 30. To see how many pieces each student will get, $\frac{300}{30} = 10$.

7. **C** Round the number of miles and number of days to numbers that are easier to divide. $\frac{7500}{15} = 500$ miles per day.

8. **C** Round each decimal to the nearest whole number and then multiply. $21(6) = 126$.

Short Response

9. **Part A.** No. Using estimation, round 19 packages to 20 and \$1.89 per package to \$2.00. If Brenda wants to buy all of the packages, 20(\$2) = \$40. Since she only has \$30, Brenda does not have enough money to buy all of the packages of plates.

   Part B. About 15 packages. Since a package costs approximately \$2, and Brenda has \$30, she could buy about 15 packages of plates. \$30 ÷ \$2 = 15.

GREATEST COMMON FACTOR AND LEAST COMMON MULTIPLE

1. **B** 16 is a factor of 32 and 48. $16(2) = 32$ and $16(3) = 48$.

2. **C** 7 is a factor of 35, 49, and 84. $7(5) = 35$, $7(7) = 49$, and $7(12) = 84$.

3. **C** Multiples of 8 are 8, 16, 24, 32, 40, ...
   Multiples of 20 are 20, 40, 60, ...
   40 is the first multiple common to both 8 and 20.

4. C Multiples of 4 are 4, 8, 12, 16, 24, 28, ...
Multiples of 8 are 8, 16, 24, 32, ...
Multiples of 12 are 12, 24, ...
24 is the first multiple common to 4, 8, and 12.

5. B 2, 7, and 8 all go into 56.
For the other choices, not all of the numbers evenly divide into 56.

6. B The size of the wall is 40×16. Multiplying these two numbers will give you the area of the wall, which is 640. An 8×8 tile has an area of 64. To see how many tiles are needed to fill the area of 640, divide $\frac{640}{64}$. Your dad needs 10 tiles to cover the wall.

7. C This is an LCM problem.
The multiples of 6 are 6, 12, 18, 24, ...
The multiples of 8 are 8, 16, 24, ...
The first day that hot dogs and tacos will be served is on day 24.

Short Response

8. Part A. This is a GCF problem because you are looking for the "greatest number of lollipops." The factors of 36 are 1, 2, 3, 4, 6, 9, 12, 18, and 26. The factors of 42 are 1, 2, 3, 6, 7, 14, 21, and 42. The GCF is 6.

Part B. You have a box of 36 lollipops and a box of 42 lollipops. In total, there are 78 lollipops. If you want 3 in each bag, $\frac{78}{3} = 26$. 26 lollipops can go into each bag.

PRIME FACTORIZATION

1. B There are different ways to solve this problem. You could make a factor tree for 36 or you could find out each answer for A, B, C, and D by multiplying.
$2^2 \times 3^2 = 4 \times 9 = 36$

2. D A prime number only has factors of 1 and itself. 13 only has factors of 1 and 13.

3. C A composite number has two or more factors. The factors of 22 are 1, 2, 11, and 22 so it is composite.

4. C You can make a factor tree or figure out each answer by multiplying.
$2^3 \times 3 \times 5 = 8 \times 3 \times 5 = 24 \times 5 = 120$

5. C $2^3 \cdot 3^2 \cdot 5 = 8 \times 9 \times 5 = 72 \times 5 = 360$

6. D $3^2 \cdot 13 = 9 \cdot 13 = 117$

Short Response

7. Part A.

This is one example of a factor tree. Any two factors that multiply to 720 can be the first branch of the tree.

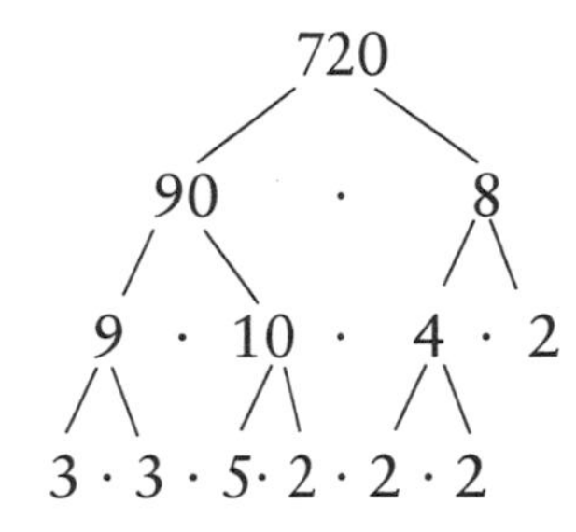

Answer = $2 \times 2 \times 2 \times 2 \times 3 \times 3 \times 5$

Part B. The answer written in exponential form is $2^4 \times 3^2 \times 5$.

8. Chris started the factor tree correctly, but he did not finish it correctly. The number 6 is not a prime number, so Chris needed to add another branch. Off of 6, he needed to break it down to 3 times 2. He would be correct if his tree had the prime factors of 2, 2, 3, and 5.

SCIENTIFIC NOTATION AND TRANSLATING SCIENTIFIC NOTATION INTO STANDARD FORM

1. C $214{,}000 \rightarrow 2.14 \times 10^5$

2. A $300 \rightarrow 3 \times 10^2$

3. C $.004 \rightarrow 4 \times 10^{-3}$

4. D $3.45 \times 10^5 \rightarrow 345{,}000$

5. C $1.25 \times 10^3 \rightarrow 1{,}250$

6. **C** There are 603 students in seventh grade.
603 written in scientific notation is 6.03×10^2.

7. **D** The first factor 23.6 is greater than or equal to 1, but not less than 10.

Short Response

8. **Part A.** Emily

Part B. Emily is correct because the first factor is greater than or equal to 1 and less than 10. Tina's first factor, 545, is greater than 1, but not less than 10.

COMPARING NUMBERS IN SCIENTIFIC NOTATION

1. **B** Write each number in standard form to compare them.

3.6×10^4	=	36,000	42,000	=	42,000
4.5×10^4	=	45,000	36,500	=	36,500

The greatest number is 45,000 (4.5×10^4).

2. **A** Since each number is written in scientific notation, look for the smallest exponent. Since 3.2×10^7 and 1.7×10^7 both have an exponent of 7, look for the smaller first factor. 1.7 is smaller than 3.2, so 1.7×10^7 is the smallest number. You may also write each number in standard form to compare which is the least.

3. **C** Since you are ordering from least to greatest, the smallest number will also have the smallest exponent. 5.3×10^2 has an exponent of 2, and there are no other exponents of 2, so this is the smallest number. Choice B has the numbers listed from smallest to largest.

4. **D** One possible method to solve this problem is to write each problem out in standard form. Order the numbers from least to greatest.

United States	3.01×10^8	=	301,000,000	#4
Canada	3.34×10^7	=	33,400,000	#1
Japan	1.275×10^8	=	127,500,000	#3
Germany	8.24×10^7	=	82,400,000	#2

5. **C** To find the smallest mass, look for the smallest exponent. Earth and Venus both have an exponent of 24. Mercury and Mars have exponents of 25.

Earth	5.98×10^{24}
Venus	4.87×10^{24}

Now compare the first factors. 4.87 is smaller than 5.98, so Venus has the smallest mass.

Short Response

6. **Part A.**

To compare the numbers, either write all of the numbers in scientific notation or write all of the numbers in standard form. This solution shows all of the numbers written in standard form.

Lake Erie	2.57×10^4	25,700	#2
Lake Huron		59,600	#4
Lake Superior	8.2×10^4	82,000	#5
Lake Michigan		57,800	#3
Lake Ontario	1.896×10^4	18,960	#1

The lakes in order from smallest to largest are:

Lake Ontario, Lake Erie, Lake Michigan, Lake Huron, Lake Superior

Part B. The Great Lake with the largest area is Lake Superior.

CHAPTER 2: ALGEBRA

EVALUATING ALGEBRAIC EXPRESSIONS

1. **C**

Rewrite the expression.	$2a + 6b - 5$
Substitute $a = 12$ and $b = 2$.	$2(12) + 6(2) - 5$
Multiply left to right.	$24 + 12 - 5$
Add.	$36 - 5$
Subtract.	31

2. C Rewrite the expression. $m^3 + n^2 - 2$
Substitute $m = 3$ and $n = 2$. $3^3 + 2^2 - 2$
Simplify the exponents. $27 + 4 - 2$
Add. $31 - 2$
Subtract. 29

3. D Rewrite the expression. $(xy)^2$
Substitute $x = 2$ and $y = 4$. $(2 \cdot 4)^2$
Multiply. 8^2
Simplify the exponent. 64

4. B Rewrite the expression. $4x^2 - y$
Substitute $x = 3$ and $y = 5$. $4(3^2) - 5$
Simplify the exponent. $4(9) - 5$
Multiply. $36 - 5$
Subtract. 31

5. C Rewrite the expression. $\frac{1}{2}(5m)(n^2)$
Substitute $m = 4$ and $n = 3$. $\frac{1}{2}(5 \cdot 4)(3^2)$
Simplify the exponent. $\frac{1}{2}(5 \cdot 4)(9)$
Multiply left to right. $10(9)$
Multiply. $\frac{1}{2}(20)(9)$
Multiply. 90

6. B Rewrite the expression. $3x + 4y$
Substitute $x = 2$ and $y = 3.5$. $3(2) + 4(3.5)$
Multiply. $6 + 14$
Add. 20

7. C Substitute the value for x and y into $5xy + 3$. $x = 3$ and $y = 1$ makes $5(3)(1) + 3 = 18$.

8. A Rewrite the formula. $\$3t + \$2p$
Substitute $t = 4$ and $p = 2$. $\$3(4) + \$2(2)$
Multiply left to right. $12 + 4$
Add. 16

9. **D** Rewrite the expression. $4g^2 + \frac{h}{3}$

Substitute $g = 2$ and $h = 12$. $4(2)^2 + \frac{12}{3}$

Simplify the exponent. $4(4) + \frac{12}{3}$

Multiply first. $16 + \frac{12}{3}$

Divide. $16 + 4$

Add. 20

Short Response

10. **Part A.** Kim is 32 years old.

Rewrite the expression.	$3t - 4$
Substitute 12 for Tommy's age.	$3(12) - 4$
Multiply.	$36 - 4$
Subtract.	32

Part B. Tommy would be 7 years old.
$3t - 4$ represents Kim's age in relation to Tommy's. If you know Kim is 17, then $3t - 4 = 17$. Solve the equation.

$$\begin{aligned} 3t - 4 &= 17 \\ +4 \quad & \quad +4 \\ \hline \frac{3t}{3} &= \frac{21}{3} \\ t &= 7 \end{aligned}$$

TRANSLATING 2-STEP VERBAL EXPRESSIONS INTO EQUATIONS

1. **C** The important concept to remember with this problem is that the two less than comes *after* six times a number.

Six times a number	$6x$
Two less than	-2
Equal to 10	$= 10$
Equation	$6x - 2 = 10$

2.	B	The sum of five and four times a number	$5 + 4x$
		equals 14	$= 14$
		Equation	$5 + 4x = 14$
3.	C	Three more than two times a number	$3 + 2n$
		is twelve	$= 12$
		Equation	$3 + 2n = 12$
4.	B	Four dollars plus thirty cents per mile	$4 + .30m$
		Your cab ride cost $36	$= 36$
		Equation	$4 + .30m = 36$
5.	A	The product of a number and twelve	$12x$
		is equal to	$=$
		the sum of thirty and	$30 +$
		twice the same number	$2x$
		Equation	$12x = 2x + 30$
6.	B	Drew charges five dollars to rake each yard	$5y$
		plus a two dollar clean-up fee	$+ 2$
		Drew made $102	$= 102$
		Equation	$5y + 2 = 102$
7.	D	Twice a number	$2x$
		plus three	$+ 3$
		is equal to	$=$
		the sum of the same number and seven	$x + 7$
		Equation	$2x + 3 = x + 7$
8.	D	The quotient of y and eight	$\frac{y}{8}$
		plus fifteen	$+ 15$
		equals thirty-one	$= 31$
		Equation	$\frac{y}{8} + 15 = 31$

Short Response

9. **Part A.**

	$39 + 0.14m$
Phone America charges $39 per month	39
Plus 0.14 per minute	$+ 0.14m$
Expression	$39 + 0.14m$

Part B.

	$40.40
Rewrite the expression.	$39 + 0.14m$
Since you are 10 minutes over, let $m = 10$.	$39 + 0.14(10)$
Multiply.	$39 + 1.40$
Add.	$40.40

SOLVING 2-STEP EQUATIONS

1. C Rewrite the equation.
Subtract 168 from both sides.
Divide both sides by 2.
Simplify.

$$2x + 168 = 432$$
$$\underline{\quad -168 \; -168}$$
$$\frac{2x}{2} = \frac{264}{2}$$
$$x = 132$$

2. D Rewrite the equation.
Add 9 to both sides.
Multiply both sides by 5.
Simplify.

$$\frac{y}{5} - 9 = 17$$
$$\underline{\quad +9 \;\; +9}$$
$$5 \cdot \frac{y}{5} = 26 \cdot 5$$
$$y = 130$$

3. C Rewrite the equation.
Subtract 12 from both sides.
Multiply both sides by 3.
Simplify.

$$\frac{m}{3} + 12 = 39$$
$$\underline{\quad -12 \;\; -12}$$
$$3 \cdot \frac{m}{3} = 27 \cdot 3$$
$$m = 81$$

4. C Rewrite the equation.
Subtract 39 from both sides.
Divide both sides by .25.
Simplify.

$$.25t + 39 = 114$$
$$\underline{\quad -39 \;\; -39}$$
$$\frac{.25t}{.25} = \frac{75}{.25}$$
$$t = 300$$

5. A Rewrite the equation.
Subtract 3 from both sides.
Divide both sides by 2.
Simplify.

$$2h + 3 = 13$$
$$\underline{\quad -3 \;\; -3}$$
$$\frac{2h}{2} = \frac{10}{2}$$
$$h = 5$$

6. B Rewrite the equation.
Subtract 5 from both sides.
Divide both sides by 4.
Simplify.

$$4n + 5 = 25$$
$$\underline{\quad -5 \;\; -5}$$
$$\frac{4n}{4} = \frac{20}{4}$$
$$n = 5$$

7. **A** Rewrite the equation. $2b + 6 = 16$
Subtract 6 from both sides. $-6 \quad -6$
Divide both sides by 2. $\frac{2b}{2} = \frac{10}{2}$
Simplify. $b = 5$

8. **D** Rewrite the equation. $\frac{x}{2} - 8 = 10$
Add 8 to both sides. $+8 \quad +8$
Multiply both sides by 2. $\frac{x}{2} = 18$
Simplify. $2 \cdot \frac{x}{2} = 18 \cdot 2$

$x = 36$

Short Response

9. **Part A.** $4s + 12 = 68$

Jenna bought four shirts at the mall.
Let s represent each shirt. $4s$
The tax was \$12. $+ 12$
For a total of \$68. $= 68$

Part B. \$14

$$4s + 12 = 68$$
$$-12 \quad -12$$
$$\frac{4s}{4} = \frac{56}{4}$$
$$s = 14$$

10. **Part A.** Anthony did not solve the problem correctly. He should have added 5 to both sides first. Instead, he tried to undo dividing by 2 by multiplying both sides by 2.

Part B.

$$\frac{x}{2} - 5 = 9$$
$$+5 \quad +5$$
$$2 \cdot \frac{x}{2} = 14 \cdot 2$$
$$x = 28$$

PROPORTIONS

1. A $\frac{\textit{cups}}{\textit{cookies}}$

$$\frac{2}{12} = \frac{x}{60}$$
$$\frac{12x}{12} = \frac{120}{12}$$
$$x = 10 \text{ cups}$$

2. B $\frac{\textit{gallons}}{\textit{cost}}$

$$\frac{10}{25} = \frac{12}{x}$$
$$\frac{10x}{10} = \frac{300}{10}$$
$$x = \$30$$

3. D $\frac{\textit{Kelly}}{\textit{Tracy}}$

$$\frac{12}{15} = \frac{20}{x}$$
$$\frac{12x}{11} = \frac{300}{12}$$
$$x = 25$$

4. C $\frac{\textit{watch}}{\textit{surveyed}}$

$$\frac{3}{8} = \frac{45}{x}$$
$$\frac{3x}{3} = \frac{360}{3}$$
$$x = 120$$

5. A $\frac{\textit{loads}}{\textit{payment}}$

$$\frac{3}{1.50} = \frac{5}{x}$$
$$\frac{3x}{3} = \frac{7.50}{3}$$
$$x = \$2.50$$

6. B $\frac{\textit{songs}}{\textit{cost}}$

$$\frac{10}{\$2} = \frac{25}{x}$$
$$\frac{10x}{10} = \frac{50}{10}$$
$$x = \$5$$

7. A $\frac{\textit{laps}}{\textit{minutes}}$

$$\frac{20}{5} = \frac{x}{20}$$
$$\frac{5x}{5} = \frac{400}{5}$$
$$x = 80$$

Short Response

8. \$22.50 $\frac{\text{cost}}{\text{pieces}}$

$$\frac{\$15}{24} = \frac{x}{36}$$
$$\frac{24x}{24} = \frac{540}{24}$$
$$x = \$22.50$$

9. **Part A.** Yes.

Method 1: Reduce each fraction $\frac{6}{18} = \frac{1}{3}, \frac{24}{72} = \frac{1}{3}$. Since both fractions reduce to $\frac{1}{3}$, they are proportional.

Method 2: Find the cross products of $\frac{6}{18}, \frac{24}{72}$; $6(72) = 432$ and $18(24) = 432$. Since the cross products are equal, the ratios form a proportion.

Part B.

Explain either Method 1 or Method 2, as stated in Part A.

SOLVING AND GRAPHING 1-STEP INEQUALITIES

1. B

$$\begin{array}{l} x + 5 \geq 12 \\ \underline{\quad -5 \quad -5} \\ x \quad\quad \geq 7 \end{array}$$

2. C

$$\begin{array}{l} y - 7 < 9 \\ \underline{\quad +7 \quad +7} \\ y \quad\quad < 16 \end{array}$$

3. D

$$2 \cdot \frac{m}{2} > 5 \cdot 2$$
$$m > 10$$

4. C

$$\frac{2d}{2} \leq \frac{20}{2}$$
$$d \leq 10$$

Any number less than or equal to 10 will work in this inequality. 11 is greater than 10, so this is not in the solution.

5. B
$$\begin{array}{l} 4 + y > 5 \\ \underline{-4 \quad -4} \\ y > 5 \end{array}$$

6. C
$$\begin{array}{l} x - 7 \le 9 \\ \underline{\ +7 \ +7} \\ x \le 16 \end{array}$$

7. A
$$\begin{array}{l} 3 + y > 7 \\ \underline{-3 \quad -3} \\ y > 4 \end{array}$$

8. D The graph shows a solution greater than or equal to 11.

Short Response

9. Part A.

$$\begin{array}{l} 3 \ge y - 4 \\ \underline{+4 \qquad +4} \\ 7 \ge y \end{array}$$

(which is the same as $y \le 7$)

Part B. 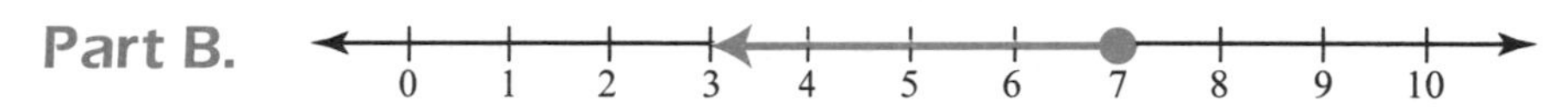

EVALUATING FORMULAS USING SURFACE AREA, RATE, AND DENSITY

1. B

Use the formula I = prt.	I = prt
Substitute p = 75, rate = 0.055, and t = 5.	I = 75(0.055)(5)
Multiply.	I = 20.625
Since it is money, round to the nearest cent.	I = \$20.63

2. B

Rewrite the formula.	$6s^2$
Substitute 3 in for s.	$6(3)^2$
Simplify the exponent.	6(9)
Multiply.	54

3. **A** Rewrite the formula. $\frac{1}{2}lwh$

Substitute $w = 4$, $h = 5$, $l = 8$. $\frac{1}{2}(4)(5)(8)$

Multiply left to right. $2(5)(8)$
$10(8)$
80

4. **A** Rewrite the formula. $d = m \div v$
Substitute $m = 725$, $v = 25$. $d = 725 \div 25$
Divide. $d = 29$

5. **B** Use the formula I = prt. I = prt
Substitute $p = 1000$, $r = 0.035$, $t = 2$. I = 1000(0.035)(2)
Multiply left to right. I = $70

6. **D** Rewrite the formula. distance = rate × time
Substitute rate and time
rate = 50 miles per hour,
time = 7.5 hours. distance = 50(7.5)
Multiply. distance = 375 miles

7. **C** Rewrite the formula. $d = m \div v$
Substitute $d = 60$ and $v = 15$. $60 = m \div 15$
An easier way to write this would be $60 = \frac{m}{15}$
Undo, divide 15 by multiplying by 15 on both sides. $15 \cdot 60 = \frac{m}{15} \cdot 15$
Simplify. $900 = m$

Short Response

8. **Part A.** $130 Rewrite the formula. I = prt
Substitute $p = 500$, $r = 0.065$, $t = 4$. I = 500(0.065)(4)
Multiply left to right. I = $130

Part B. $630 To find the total amount in the savings account, add the amount deposited and the amount of interest earned in 4 years.
500 + 130 = $630

CHAPTER 3: GEOMETRY

PLOTTING POINTS ON THE COORDINATE PLANE

1. **C** The point (–3,–6) means from the origin move 3 units to the left and 6 units down.
2. **A** The point (0, 3) means from the origin move 0 units left or right and 3 units up.
3. **B** The coordinates (–3, 0), (–1, –3), (1, 3), and (2, –1) are on the graph.
4. **D** The points (–4, 3), (0, 0), and (0, –3) are on the graph.
5. **C** From the origin, move 3 units left and 3 units up. The point (–3, 3) will complete the graph.
6. **D** The point (4, –2) is not on the graph.

Short Response

7. **Part A.**

See graph below.

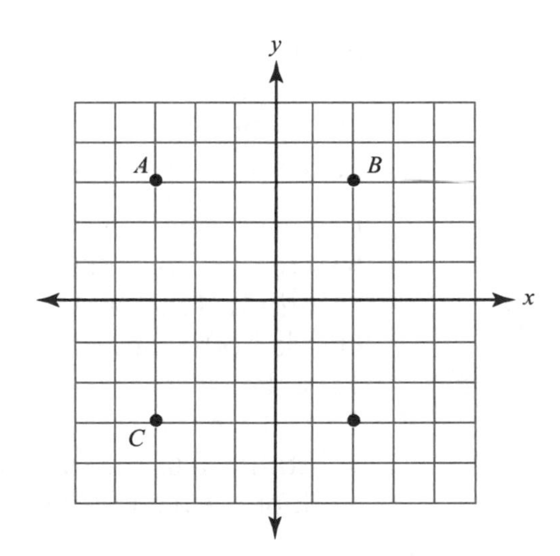

Part B. The point (2, –3) will complete the graph to make the figure a rectangle.

AREA ON THE COORDINATE PLANE

*Note: When dividing figures you can draw horizontal and/or vertical lines. Solutions here give one method.

1. B Count the squares inside the figure.
2. B Count the squares inside the figure.
3. D Draw a vertical line at (–4,0) and (–1,0).
4. B Draw a vertical line at (–5,0) and (–3,0).
5. C P has an area of 10, *M*, *N*, and *R* have an area of 8.
6. B There are 7 squares inside figure *S*.

Short Response

7. Answer will vary. The figure must have 42 squares inside of it.

AREA AND CIRCUMFERENCE

1. B Use the formula $C = \pi d$

Rewrite the formula	$C = \pi d$
Substitute 3.14 for π and 62.8 for C	$62.8 = 3.14d$
Divide both sides by 3.14	$20 = d$

2. A Use the formula $A = \pi r^2$

Rewrite the formula	$A = \pi r^2$
Substitute 28.26 for A and 3.14 for π	$28.26 = 3.14r^2$
Divide both sides by 3.14	$9 = r^2$
Take the square root of both sides	$3 = r$

3. C Use the formula $A = \pi r^2$

Rewrite the formula	$A = \pi r^2$
Substitute 49π for A	$49\pi = \pi r^2$
Divide both sides by π	$49 = r^2$
Take the square root of both sides	$7 = r$
To find circumference use, $C = 2\pi r$	$C = 2\pi r$
Substitute 7 in for the radius	$C = 2(\pi)(7)$
Multiply, leave in terms of π	$C = 14\pi$

4. D

Use $C = 2\pi r$ to find the radius	$C = 2\pi r$
Substitute 60π in for C	$60\pi = 2\pi r$
Divide both sides by 2π	$30 = r$
Use $A = \pi r^2$ to find the area	$A = \pi r^2$
Substitute 30 in for *r*, leave π	$A = \pi(30)^2$
Simplify	$A = 900\pi$

5. **B** Use $C = 2\pi r$ since circumference is given. $C = 2\pi r$
   Substitute 12π in for C $12\pi = 2\pi r$
   Divide both sides by 2π $6 = r$

6. **C** Use $C = 2\pi r$ since circumference is given. $C = 2\pi r$
   Substitute 36π in for C $36\pi = 2\pi r$
   Divide both sides by 2π $18 = r$
   Use $A = \pi r^2$ $A = \pi r^2$
   Substitute 18 for r, leave π $A = \pi (18)^2$
   Simplify $A = 324\pi$

7. **A** The circumference of the first circle is 10π and the circumference of the second circle is 12π. The difference is 2π.

Short Response

8. Find the radii for the red and the blue circles.

Red Circle	Blue Circle
$A = \pi r^2$	$C = 2\pi r$
$16\pi = \pi r^2$	$12\pi = 2\pi r$
$16 = r^2$	$6 = r$
$4 = r$	

The blue circle has a larger radius.

MISSING ANGLES IN QUADRILATERALS

1. **A** $85° + 120° + 110° = 315°$
   $360° - 315° = 45°$

2. **C** $65° + 95° = 160°$
   $360° - 160° = 200°$
   Since the two remaining angles are equal, divide 200° by 2. Each angle is 100°.

3. **D** All four angles must add up to 360°.
   $70° + 80° + 90° + 120° = 360°$

4. **C** All four angles must add up to 360°. Two of the angles already add to 140°.
   $360° - 140° = 220°$
   Since the last two angles are equal,
   $\frac{220}{2} = 110°.$

5. **D** All four angles must add up to 360°.
$80° + 80° + 150° + 50° = 360°$

6. **A** All four angles must add up to 360°.
$30° + 70° + 110° + 150° = 360°$.

7. **B** The angles in Choices A, C, and D each add to 360°. For Choice B the angles add to 325°. Since the angles in a quadrilateral must equal 360°, these angles do not work.

Short Response

8. **Part A.** There are many possible solutions. All four angles must be different and must add up to 360°.

Part B. There are many possible solutions. Two of the angles must be equal, the other two angles unequal. The angles must add up to 360°.

Part C. Yes, the angles in a quadrilateral can all be equal. The angles could each be 90°, like in a square or a rectangle.

FACES AND BASES OF 3D SHAPES

1. **B** The 6 faces are the front, back, top, bottom, and 2 sides.
2. **A** The vertices are the corners. There are 4 corners on the top and 4 corners on the bottom.
3. **B** This is a rectangular prism.
4. **B** The 5 faces are the top, bottom, and 3 sides.
5. **B** There are 2 bases; they are both triangles.
6. **A** The edges are where the faces meet. There is a top edge, 3 edges on one side, 3 edges on the other side, and 2 on the bottom.
7. **A** The figure is a triangular prism; it is named so because of its bases.
8. **C** The 4 faces are the 3 sides and the bottom.
9. **C** There are 4 points that are the vertices.
10. **D** Since the base and the sides are triangles, this is a triangular pyramid.

Short Response

11.

Name of solid	Rectangular Pyramid
Number of faces	5
Number of edges	8
Number of vertices	5

VOLUME OF PRISMS AND CYLINDERS

1.	B	Use the formula $V = lwh$	$V = lwh$
		Substitute $l = 5.5$, $w = 2.5$, $h = 6$	$V = (5.5)(2.5)(6)$
		Multiply	$V = 82.5$ ft^3
2.	C	Use the formula $V = \pi r^2 h$	$V = \pi r^2 h$
		Substitute $\pi = 3.14$, $r = 3$, $h = 5$	$V = 3.14(3^2)(5)$
		Follow the order of operations	$V = 141.3$ in^3
3.	C	Use the formula $V = lwh$	$V = lwh$
		Substitute $V = 432$, $L = 8$, $w = 6$	$432 = 8(6)(h)$
		Multiply	$432 = 48h$
		Divide	$h = 9$ in.
4.	A	Use the formula $V = \pi r^2 h$	$V = \pi r^2 h$
		Substitute $V = 942$, $\pi = 3.14$, $r = 5$	$942 = 3.14(5^2)(h)$
		Simplify the exponent	$942 = 3.14(25)(h)$
		Multiply	$942 = 78.5h$
		Divide	$h = 12$ in.
5.	A	Use the formula $V = \pi r^2 h$	$V = \pi r^2 h$
		Substitute $V = 301.44$, $\pi = 3.14$, $r = 4$	$301.44 = 3.14(4^2)(h)$
		Simplify the exponent	$301.44 = 3.14(16)(h)$
		Multiply	$301.44 = 50.24h$
		Divide	$h = 6$ ft.
6.	B	Use the formula $V = lwh$	$V = lwh$
		Substitute $V = 2340$, $l = 15$, $h = 13$	$2340 = 15(w)13)$
		Multiply	$2340 = 195w$
		Divide	$w = 12$ cm
7.	C	Use the formula $V = \pi r^2 h$	$V = \pi r^2 h$
		Substitute $r = 3$, $h = 10$	$V = \pi\ (3^2)(10)$
		Simplify the exponent	$V = \pi(9)(10)$
		Multiply	$V = 90\pi$

8. A Use the formula $V = \pi r^2 h$ | $V = \pi r^2 h$
Substitute $V = 339.12$, $\pi = 3.14$, $h = 12$ | $339.12 = 3.14\,(r^2)(12)$
Multiply | $339.12 = 37.68r^2$
Divide | $9 = r^2$
Take the square root of both sides | $r = 3$ cm

9. C $V = lwh$. Multiply each of the choices to see which three numbers will equal 630. $7(9)(10) = 630$.

10. D $V = \pi r^2 h$. Substitute each of the choices in for r and h to see which set of numbers will make the $V = 75\pi$. When the radius = 5 and the height = 3, $V = 5^2(3)\pi = 75\pi$.

Short Response

11. Use the formula $V = \pi r^2 h$ | $V = \pi r^2 h$
Substitute $V = 1582.56$, $\pi = 3.14$, $h = 14$ | $1582.56 = 3.14\,(r^2)(14)$
Multiply | $1582.56 = 43.96r^2$
Divide | $36 = r^2$
Take the square root of both sides | $6 = r$

Since the radius = 6, the diameter = 2(6) = 12 feet.

SURFACE AREA OF PRISMS AND CYLINDERS

1. B Use the formula $SA = 2\pi rh + 2\pi r^2$ | $SA = 2\pi rh + 2\pi r^2$
Substitute $\pi = 3.14$, $r = 3$, $h = 5$ | $SA = 2(3.14)(3)(5) + 2(3.14)(3^2)$
Follow the order of operations | $SA = 94.2 + 56.52$
Add | $SA = 150.72$ sq. cm

2. B Use the formula $SA = 2wl + 2lh + 2wh$ | $SA = 2wl + 2lh + 2wh$
Substitute $w = 3$, $l = 6$, $h = 4$ | $SA = 2(3)(6) + 2(6)(4) + 2(3)(4)$
Follow the order of operations | $SA = 36 + 48 + 24$
Add | $SA = 108$ sq. cm

3. C Use the formula $SA = 2\pi rh + 2\pi r^2$ | $SA = 2\pi rh + 2\pi r^2$
Substitute $\pi = 3.14$, $r = 2$, $h = 6$ | $SA = 2(3.14)(2)(6) + 2(3.14)(2^2)$
Follow the order of operations | $SA = 75.36 + 25.12$
Add | $SA = 100.48$
Round to nearest tenth | $SA = 100.5$ in^2

4. D Use the formula $SA = 2wl + 2lh + 2wh$ — $SA = 2wl + 2lh + 2wh$

Substitute $w = 4, l = 8, h = 5$ — $SA = 2(4)(8) + 2(8)(5) + 2(4)(5)$

Follow the order of operations — $SA = 64 + 80 + 40$

Add — $SA = 184 \text{ in}^2$

5. B Use the formula $SA = 2\pi rh + 2\pi r^2$ — $SA = 2\pi rh + 2\pi r^2$

Substitute $r = 12, h = 30$ — $SA = 2(\pi)(12)(30) + 2(\pi)(12^2)$

Follow the order of operations — $SA = 4320\pi + 288\pi$

Add — $SA = 4608\pi$ sq. m

6. C Since a vase is open at the top, use the formula $SA = 2\pi rh + \pi r^2$ instead of $2\pi rh + 2\pi r^2$.

Use the formula $SA = 2\pi rh + \pi r^2$ — $SA = 2\pi rh + \pi r^2$

Substitute $\pi = 3.14, r = 4, h = 7$ — $SA = 2(3.14)(4)(7) + (3.14)(4^2)$

Follow the order of operations — $SA = 175.84 + 50.24$

Add — $SA = 226.08$ sq. in.

$SA = 6s^2$

7. C Use the formula $SA = 6s^2$ — $SA = 6(7^2)$

Simplify the exponent — $SA = 6(49)$

Multiply — $SA = 294$ sq. m

8. A Use the formula $SA = wh + lw + lh + ls$ — $SA = wh + lw + lh + ls$

Substitute $l = 10, w = 4, h = 3, s = 5$ — $SA = 4(3) + 10(4) + 10(3) + 10(5)$

Follow the order of operations — $SA = 12 + 40 + 30 + 50$

Add — $SA = 132 \text{ in}^2$

Short Response

9. Find the surface area of the original box and the new box.

Original Box

Use the formula $SA = 2wl + 2lh + 2wh$ — $SA = 2wl + 2lh + 2wh$

Substitute $w = 6, l = 8, h = 10$ — $SA = 2(6)(8) + 2(8)(10) + 2(6)(10)$

Follow the order of operations — $SA = 96 + 160 + 120$

Add — $SA = 376$ sq. in.

New Box

Use the formula $SA = 2wl + 2lh + 2wh$ — $SA = 2wl + 2lh + 2wh$

Substitute $w = 4$, $l = 6$, $h = 8$ — $SA = 2(4)(6) + 2(6)(8) + 2(4)(8)$

Follow the order of operations — $SA = 48 + 96 + 64$

Add — $SA = 208$ sq. in.

To find how much less material is needed to make the new box, subtract the SA of the new box from the SA of the original box. $376 - 208 = 168$ square inches.

CHAPTER 4: MEASUREMENT

CONVERTING MASS AND WEIGHT

1. **D** Tons is a customary unit. This is a good measurement for objects with a large mass, since there are 2000 pounds in a ton.
2. **A** Meters is the metric unit used to measure length.
3. **A** On the customary staircase, you move 1 step to the right. There are 16 pounds in an ounce, so multiply by 16.
4. **C** On the metric staircase, move 1 step to the left, so divide by 10.
5. **A** On the metric staircase, move 3 steps to the left, so divide by 1000.
6. **B** On the customary staircase, move 1 step to the right, so multiply by 16. 3 times 16 equals 48.
7. **C** On the metric staircase, move 3 steps to the right, so multiply by 1000.
8. **C** On the metric staircase, move 3 steps to the left, so divide by 1000.
9. **B** On the customary staircase, move 1 step to the left, so divide by 2000.
10. **B** To solve this problem, you need to make the units the same.

    Solution 1: Change 6 meters to centimeters. On the metric staircase, move 2 steps to the right. Multiply by 100. 6m equals 600 cm.
    60 cm < 600 cm

Solution 2: Change 60 cm to meters. On the metric staircase, move 2 steps to the left. Divide by 100. 60 cm equals 0.6 meters.
0.6 meters < 6 meters

11. A *Solution 1:* Change 5 grams to milligrams. On the metric staircase, move 3 steps to the right. Multiply by 1000. 5 grams equals 5000 milligrams.
5000 mg > 512 mg

Solution 2: Change 512 milligrams to grams. On the metric staircase, move 3 steps to the left. Divide by 1000. 512 mg equals 0.512 grams.
5g > 0.512 grams

Short Repsonse

12. Part A

To compare the measurements, all of the measurements must be converted to the same unit. There is more than one correct way to solve this problem. The solution listed here changed all the units to meters.

1200 centimeters Divide by 100 = 12 meters #4	8000 millimeters Divide by 1000 = 8 meters #3	35 millimeters Divide by 1000 = 0.035 meters #1
1 kilometer Multiply by 1000 = 1000 meters #6	2 meters Stays the same = 2 meters #2	30 meters Stays the same = 30 meters #5

Order from least to greatest. When writing the answer, write the original problem.

Final Answer:
35 millimeters, 2 meters, 8000 millimeters, 1200 centimeters, 30 meters, 1 kilometer.

Part B.

Solution 1: Eight pounds would not be an appropriate measurement to add because all of the examples the teacher listed are metric units and pounds is a customary unit.

Solution 2: Eight pounds would not be an appropriate measurement to add because all of the examples the teacher listed are units of length and pounds is a unit of weight.

CONVERTING CAPACITY AND VOLUME

1. **A** On metric staircase, move 3 steps to the left, so divide by 1000.
2. **D** On the customary staircase, move 1 step to the right, so multiply by 8.
3. **B** On the customary staircase, move 1 step to the left, so divide by 4.
4. **B** On the customary staircase, move 1 step to the right, so multiply by 2.
5. **D** On the metric staircase, move 3 steps to the right, so multiply by 1000.
6. **C** On the customary staircase, move 1 step to the right, so multiply by 2.
7. **C** On the customary staircase, move 2 steps to the right, so multiply by 4.
8. **C** On the customary staircase, move 1 step to the right, so multiply by 4.
9. **D** On the metric staircase, move 3 steps to the right, so multiply by 1000.
10. **B** The basic unit of mass in the metric system is the gram.

Short Response

11. **Part A.**

On the metric staircase, move 3 steps to the left, so divide by 1000.
12.5 liters = 0.125 kiloliters

Part B.

You divide when you are changing from liters to kiloliters because you are changing from a smaller unit to a larger unit and you need less of the larger unit.

CHOOSING THE APPROPRIATE TOOL AND JUSTIFYING THE REASONABLENESS OF THE MASS OF AN OBJECT

1. A When comparing two objects (cell phone and MP3 player), use a balance.
2. D Since a tractor is a very large object, use a truck scale.
3. A When comparing two objects, use a balance.
4. B 1 box = 10 grams
   8 boxes = 8(10) = 80 grams
5. B If you and the puppy weigh 110 pounds, when you stand on the scale without the puppy you will weigh less. 90 pounds is the best estimate of your weight.

Short Response

6. 1 pack of gum = 5 grams. The MP3 players weighs 35 grams.

   $\frac{35}{5} = 7$ packs of gum equals the mass of the MP3 player.

   Since there are 5 pieces of gum in a pack, 5(7) = 35.
   The mass of the MP3 player is equal to 35 pieces of gum.

ESTIMATING SURFACE AREA

1. C $SA = 6s^2$
   $SA = 6(5^2)$
   $SA = 6(25)$
   $SA = 150 \text{ cm}^2$
2. A $SA = 2wl + 2lh + 2wh$
   $SA = 2(1)(2) + 2(2)(2) + 2(1)(2)$
   $SA = 4 + 8 + 4$
   $SA = 16 \text{ in}^2$
3. B $SA = 2\pi rh + 2\pi r^2$
   $SA = 2(3)(4)(8) + 2(3)(4^2)$
   $SA = 192 + 96$
   $SA = 288 \text{ cm}^2$
4. D $SA = 6s^2$
   $SA = 6(4^2)$
   $SA = 6(16)$
   SA = 96 sq. in.

5. C $SA = wh + lw + lh + ls$
$SA = 2(7) + 7(2) + 7(4) + 7(5)$
$SA = 14 + 14 + 28 + 35$
$SA = 91$ sq. cm

6. B $SA = 6s^2$
$SA = 6(12^2)$
$SA = 6(144)$
$SA = 864 \text{ in}^2$

7. B $SA = 2\pi rh + 2\pi r^2$
$SA = 2(3)(3)(10) + 2(3)(3^2)$
$SA = 180 + 54$
$SA = 234$ sq. in.

Short Response

8. **Part A.**
Since a vase has an open top, use the formula
$SA = wl + 2lh + 2wh$

$SA = wl + 2lh + 2wh$
$SA = 3(11) + 2(5)(11) + 2(3)(11)$
$SA = 33 + 110 + 66$
$SA = 209$ sq. in.

Part B.
The formula $SA = wl + 2lh + 2wh$ is used because a vase has an open top. Therefore, you do not need to add in the surface area of the top. You only need to add in the surface area of the bottom.

CONSTRUCTING CENTRAL ANGLES

1. C $\angle CDE$ passes through the center of the circle.
2. B $\angle GHJ$ passes through the center of the circle.
3. B The angles in a triangle add up to 180°. The two angles add up to 150°.
$180 - 150 = 30$
This angle also passes through the center of the circle.
4. B The angles in a triangle add up to 180°. The two angles add up to 120.
$180 - 120 = 60$
This angle also passes through the center of the circle.
5. D Choice D does not pass through the center of the circle.

Short Response

6. Position your protractor correctly on the radius. Start from 0° on the protractor and measure to 45°. Mark where 45° would be on the circle. Using a ruler or your protractor, draw the new radius.

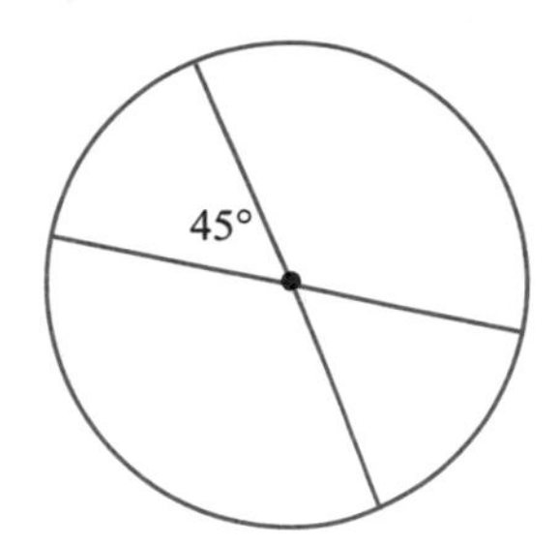

RELATIVE ERROR

1. **A** $\frac{0.4}{10} = 0.04 = \frac{4}{100} = \frac{1}{25}$

2. **A** 7.2 – 0.4 = 6.8. 7.2 + 0.4 = 7.6 The range is 6.8 lbs. to 7.6 lbs.

3. **D** $\frac{2}{8\frac{1}{2}} = \frac{2}{\left(\frac{17}{2}\right)}$
$= 2 \div \frac{17}{2}$
$= 2 \cdot \frac{2}{14}$
$= \frac{4}{17}$

4. **C** 5.4 – .25 = 5.15. 5.4 + .25 = 5.65. The range is 5.15 in. to 5.65 in.

5. **C** 5 – 0.4 = 4.6 and 5 + 0.4 = 5.4 lbs.

6. **B** $\frac{0.2}{10} = 0.02 = \frac{2}{100} = \frac{1}{50}$

Short Response

7. **Part A.**

   1.5 – .25 = 1.25
   1.5 + .25 = 1.75
   The range is 1.25 lbs. to 1.75 lbs.

   Part B.

$$\frac{.25}{1.5} = \frac{\left(\frac{1}{4}\right)}{\left(\frac{3}{2}\right)}$$
$$= \frac{1}{4} \div \frac{3}{2}$$
$$= \frac{1}{4} \cdot \frac{2}{3}$$
$$= \frac{2}{12} = \frac{1}{6}$$

   The relative error is $\frac{1}{6}$.

CHAPTER 5: PROBABILITY

LISTING OUTCOMES OF COMPOUND EVENTS

1. **D**

   A B C D
   H T H T H T H T

   The outcomes are: AH, AT, BH, BT, CH, CT, DH, DT.
   There are 8 possible outcomes.

2. **C**

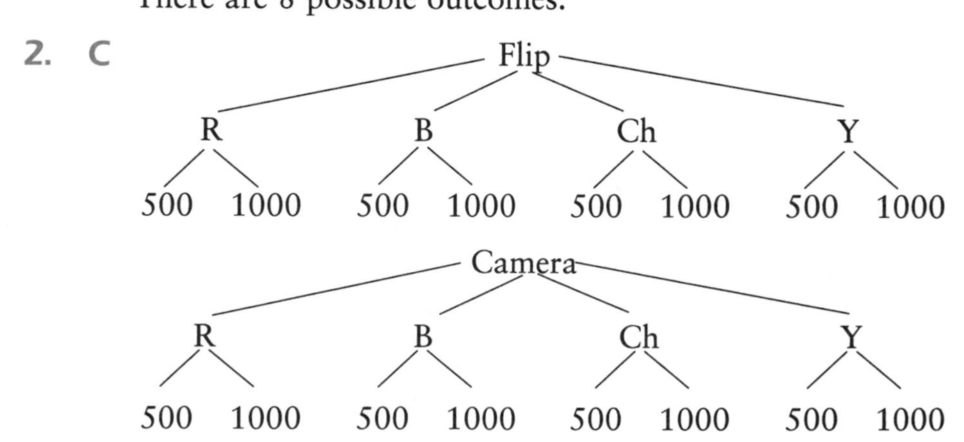

The outcomes are F-R-500, F-R-1000, F-B-500, F-B-1000, F-Ch-500, F-Ch-1000, F-Y-500, F-Y-1000, C-R-500, C-R-1000, C-B-500, C-B-1000, C-Ch-500, C-Ch-1000, C-Y-500, C-Y-1000. There are 16 possible outcomes.

3. **B**

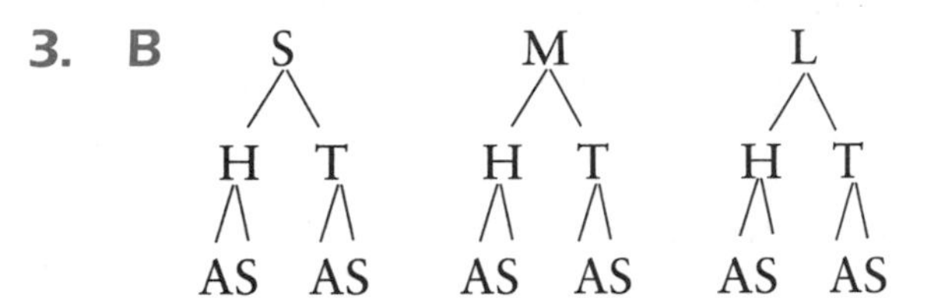

The outcomes are SHA, SHS, STA, STS, MHA, MHS, MTA, MTS, LHA, LHS, LTA, LTS. There are 12 possible outcomes.

4. Make a tree diagram.

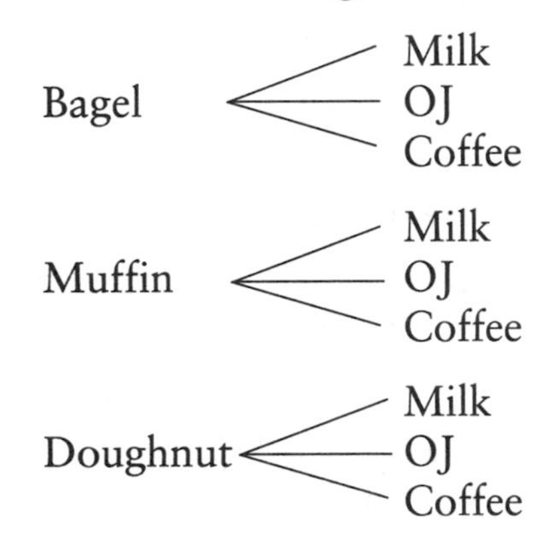

The 9 possible outcomes are BM, BO, BC, MM, MO, MC, DM, DO, DC.

5. Make a tree diagram.

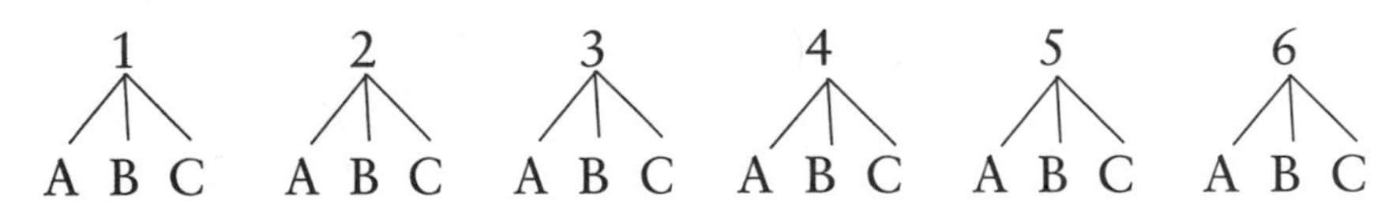

The 18 possible outcomes are 1A, 1B, 1C, 2A, 2B, 2C, 3A, 3B, 3C, 4A, 4B, 4C, 5A, 5B, 5C, 6A, 6B, 6C.

6. Make a tree diagram.

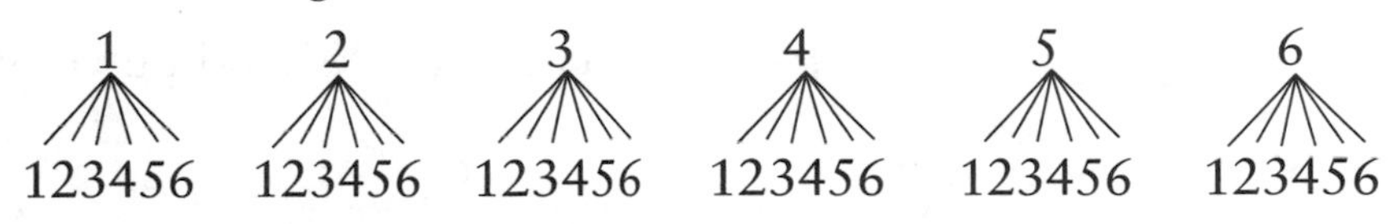

There are 36 possible outcomes.

7. Make a tree diagram.

The 12 possible outcomes are X1, X2, X3, X4, Y1, Y2, Y3, Y4, Z1, Z2, Z3, Z4.

8. Make a tree diagram.

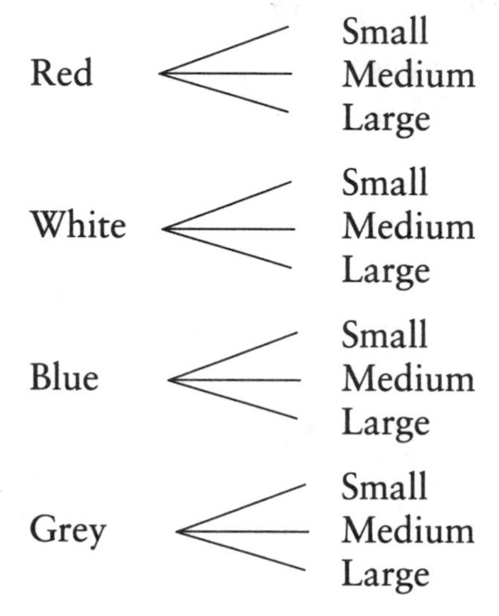

The 12 possible outcomes are RS, RM, RL, WS, WM, WL, BS, BM, BL, GS, GM, GL.

Short Response

9. **Part A.** Make a tree diagram.

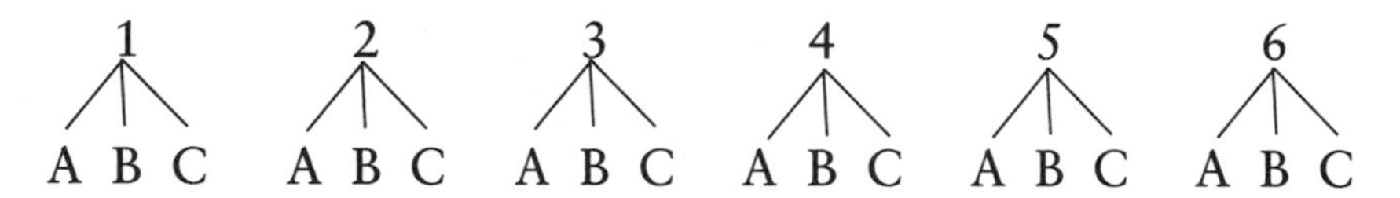

The possible outcomes are:

1A, 1B, 1C, 2A, 2B, 2C, 3A, 3B, 3C, 4A, 4B, 4C, 5A, 5B, 5C, 6A, 6B, 6C

Part B. There are 18 total outcomes.

THE FUNDAMENTAL COUNTING PRINCIPLE

1. **D** There are 6 outcomes when you roll the die the first time and 6 outcomes when you roll the die the second time. 6(6) = 36 outcomes.

2. **D** Since the code has to be four digits long and the digits 0–9 can be used, there are 10 choices for each spot. Therefore, 10(10)(10)(10) = 10,000 possible security codes.

3. **C** The trick here is to remember that each coin has 2 outcomes. Since there are five coins, the number of outcomes is 2(2)(2)(2)(2) = 32 outcomes.

4. **D** There are three choices for appetizers, four choices for main dishes, and two choices for dessert. Therefore, 3(4)(2) = 24 possible outcomes.

5. **D** There are 2 outcomes for the penny, 2 outcomes for the dime, 2 outcomes for the nickel, and 6 outcomes for the die. Therefore, 2(2)(2)(6) = 48 possible outcomes.

6. **C** For the registration sticker, consonants and digits can be repeated. There are 21 choices for the first consonant spot and 21 choices for the second consonant spot. There are 10 choices for the third, fourth, fifth, and sixth spots for the digits. Therefore, 21(21)(10)(10)(10)(10) = 4,410,000 possible outcomes.

7. **C** There are two choices for the first spot and since the letter has to be a vowel, there are seven letters left for the second spot, seven letters for the third spot, and two vowels to choose from for the last spot. Therefore, 2(7)(7)(2) = 196.

8. **B** There are four outcomes on the spinner and six outcomes on the cube, so 4(6) = 24 possible outcomes.

Short Response

9. There are six possible outcomes every time you roll the die. Since you are rolling the die four times, there are 6(6)(6)(6) = 1,296 outcomes.

10. **Part A.**

There are three choices of ice cream, three types of cones, and two choices for sprinkles. Therefore, 3(3)(2) = 18 possible outcomes.

Part B.

There are three choices of ice cream, three types of cones, and two choices for sprinkles. You multiply to find the number of outcomes, therefore, 3(3)(2) = 18.

PROBABILITY OF DEPENDENT EVENTS

1. A $\frac{1}{2}\cdot\frac{3}{6}=\frac{1}{2}\cdot\frac{1}{2}=\frac{1}{4} \quad \frac{1}{4}=0.25$

2. B $\frac{15}{23}\cdot\frac{8}{22}=\frac{120}{506}=\frac{60}{253}$

3. C $\frac{26}{52}\cdot\frac{25}{51}$

4. C $\frac{4}{21}\cdot\frac{5}{20}=\frac{4}{21}\cdot\frac{1}{4}=\frac{1}{21}$

5. B $\frac{9}{15}\cdot\frac{8}{14}=\frac{3}{5}\cdot\frac{4}{7}=\frac{12}{35}$

6. A $\frac{3}{7}\cdot\frac{1}{6}=\frac{3}{42}=\frac{1}{14}$

7. D $\frac{10}{20}\cdot\frac{6}{19}=\frac{1}{2}\cdot\frac{6}{19}=\frac{3}{19}$

8. B $\frac{4}{20}\cdot\frac{2}{19}=\frac{1}{5}\cdot\frac{2}{19}=\frac{2}{95}$

Short Response

9. **Part A.** $\frac{39}{95}$

$\frac{13}{20}\cdot\frac{12}{19}=\frac{39}{95}$

Part B. Yes, the probability of choosing a letter that is not a vowel affects the probability of choosing another letter that is not a vowel because after the first letter is drawn there are only 19 letters left. This first pick will affect what you can pick on the second draw.

10. Part A. The first mistake is that there are 10 cards, not 9, because zero is included. The second mistake is that after James picked the first card, since it is not replaced, there were only 8 cards left, not 9.

Part B.

$$\frac{5}{10} \cdot \frac{4}{9} = \frac{20}{90} = \frac{2}{9}$$

EXPERIMENTAL PROBABILITY

1. **B** $\frac{6}{20} = \frac{3}{10}$

2. **B** $\frac{13}{100} = \frac{x}{200}$

   $\frac{100x}{100} = \frac{2600}{100}$

   $x = 26$

3. **A** $\frac{25}{100} = \frac{1}{4}$

4. **D** Twenty out of 100 students watch *CSI*. Find out how many this would be out of 500.

   $\frac{20}{100} = \frac{x}{500}$

   $\frac{100x}{100} = \frac{10000}{100}$

   $x = 100$

5. **B** $\frac{7}{40}$ The spinner was spun 40 times. Frequency tells you how many times the 3 occurred.

6. **B** Theoretical probability is probability of what should happen. If there are five equal sectors, there is a 1 out of 5 chance of getting a 3.

7. **C** $\frac{6}{40} = \frac{x}{400}$

$\frac{40x}{40} = \frac{2400}{40}$

$x = 60$

8. **C** $\frac{15}{20} = \frac{x}{300}$

$\frac{20x}{20} = \frac{4500}{20}$

$x = 225$

9. **D** If 1 out of 4 students own an iPod, 3 out of 4 do not own an iPod.

$\frac{3}{4} = \frac{x}{500}$

$\frac{4x}{4} = \frac{1500}{4}$

$x = 375$

Short Response

10. **Part A.** Amy.
Amy had the highest number of tails.

Part B. 32
Possible solutions:

$\frac{8}{20} = \frac{x}{80}$

$\frac{20x}{20} = \frac{640}{20}$

$x = 32$

or

$\frac{8}{20} = \frac{x}{80}$

20 times 4 will give you 80, so 8 times 4 = 32.

CHAPTER 6: STATISTICS

FREQUENCY TABLES

1. **D** Add up the number of students. $10 + 5 + 2 + 3 = 20$

2. **C** Baseball had 2 votes out of 20.
   Set up a proportion to see how much this is out of 100.

   $$\frac{2}{20} = \frac{x}{100}$$
   $$\frac{20x}{20} = \frac{200}{20}$$
   $$x = 10$$

   10% of the votes were for baseball.

3. **B** Find the sum for the two sports over the total number of students.
   Set this equal to x over 100, and solve the proportion to find the percent.

   $$\frac{15}{20} = \frac{x}{100}$$
   $$\frac{20x}{20} = \frac{1500}{20}$$
   $$x = 75\%$$

4. Completed table:

Color of Skittle	Percent
Brown	30%
Blue	20%
Green	10%
Orange	10%
Yellow	30%

5. C If there are 30 total Skittles, set up a proportion to see how many are green.

$$\frac{x}{30} = \frac{10}{100}$$
$$\frac{100x}{100} = \frac{300}{100}$$
$$x = 3$$

6. C Brown and blue add to 50%.

7. C All of the statements are true except for C. This statement is not a fact based on the table.

8. Completed table:

Number of Pets	Tally	Frequency
0	||||	4
1	~~||||~~ |||	8
2	|||	3
3	||	2
4	|	1

9. C Add up the frequency column: $4 + 8 + 3 + 2 + 1 = 18$.

10. B One pet owned has the highest frequency.

Short Response

11. **Part A.**

Find the total number of each color marble:
Red = $5 + 8 + 2 = 15$
Blue = $3 + 2 + 11 = 16$
White = $4 + 8 + 1 = 13$
Make the frequency table showing each color and the total number for each color.

Color	Total Number
Red	15
Blue	16
White	13

Part B.
I do not agree that when combined white seems to be the most popular color. White had the smallest frequency, so it cannot be the most popular. The highest frequency was blue. Therefore, blue is the most popular color.

VENN DIAGRAMS

1. **D** Add up all of the numbers in the country circle:
   6 + 3 + 2 + 1 = 12
2. **D** Look for the number only in rock, not overlapping any other circle. Eight students like only rock.
3. **B** Look where all three circles overlap. Two students like all three types of music.
4. **A** Look where the country and pop circles overlap. Only one students likes both.
5. **B** Add up all of the numbers in the circles:
   6 + 3 + 2 + 1 + 8 + 1 + 4 = 25
   Since 30 students were surveyed, 5 students must listen to some other type of music.

6. Completed Venn diagram:

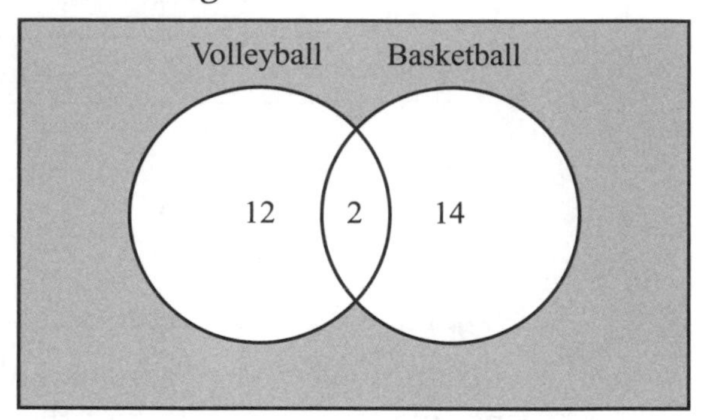

7. **B** Look at the number only in the volleyball circle. 12 students signed up.

8. **C** Look at the number only in the basketball circle. 14 students signed up.

9. **A** Add up the numbers in the circles: 12 + 2 + 14 = 28. Since there were 30 students total, 2 students must have signed up for another sport.

Short Response

10. **Part A.**

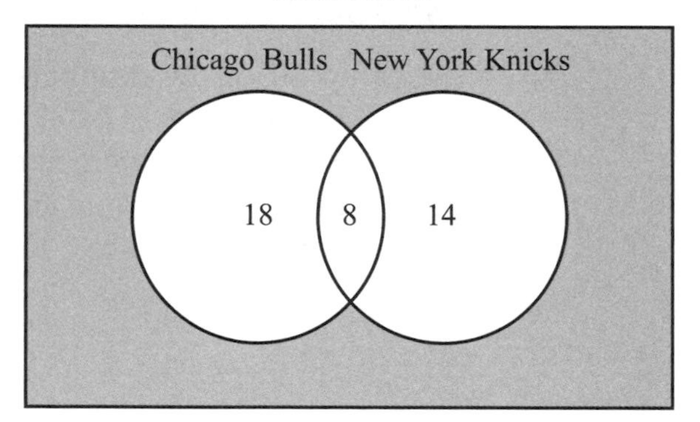

Part B.

14 people have been to a New York Knicks game only.

COLLECTING, READING, AND INTERPRETING DATA

1. **C** Each day of the week is a category, so a bar graph is best.
2. **D** The data given is in intervals 0–2, 3–5, 6–8, and 9–11, so a histogram is best.
3. **C** A line graph shows a trend over a period of time.
4. **B** Each circle represents 2 points and Molly has 2.5 circles: $2(2.5) = 5$
5. **D** There are 8.5 circles in all: $8.5(2) = 13.5$
6. **C** Look at August to see that the point is at 9 pounds.
7. **B** Dec = 12 pounds and Sept = 8 pounds: $12 - 8 = 4$
8. **C** Between 60–64, 45 people retired; for 50–54, 10 people retired: $45 - 10 = 35$
9. **C** Add up the heights of the bars for 55–59, 60–64, 65–69: $25 + 25 + 40 = 110$
10. **B** Add up the heights of the bars for 65–69 and 70–74: $40 + 5 = 45$

Short Response

11. **Part A.** A bar graph.

    Part B. The field trip activities are categories, not numbers, so a bar graph is appropriate. The height of each bar will show how many students are in favor of the activity.

RANGE

1. **C** $6.40 – $2.10 = $4.30
2. **C** $75 - 5 = 70$
3. **B** $99 - 68 = 31$
4. **B** Fill in what you know:
   High – low = range. High – 62 = 25. Since we do not know the high score, add 25 to the 62 to get the high score: $62 + 25 = 87$. The high score is 87.
   Scott scored the second highest score by 8 points. If the high score is 87, Scott scored 79 ($87 - 8 = 79$).

5. **D** $28 - (-6) = 34$
6. **C** High – low = range. Plug in what you know: $92 - L = 42$. The low score must be 50. If Kevin beat the low score by 2 points, he did 2 points better than 50. So Kevin got a 52.
7. **D** $102 - 49 = 53$
8. **C** $8 - 4 = 4$

Short Response

9. **Part A.**

   The highest number is 52 and the lowest is 3: $52 - 3 = 49$

   Part B.

   To find the range, you find the difference between the highest and lowest numbers. The highest number is 52 and the lowest number is 3: 52 minus 3 equals 49. So the range is 49.

MEASURES OF CENTRAL TENDENCY

1. **C** Find the mean:

$$\frac{3.2 + 4.1 + 6.0 + 0.4 + 1.2 + 1.8 + 0.9 + 1.8}{8}$$

$$\frac{19.4}{8}$$

$$2.425 = 2.4$$

2. **B** Put the numbers in order from least to greatest: 52, 53, 55, 56, 58, 60, 60, 61. The middle is between 56 and 58: $(56 + 58) / 2 = 57$
3. **D** The mean = $(16 + 28 + 10 + 11 + 23) / 5 = 17.6$
   Put the numbers in numerical order to find the median.
   The median is 16.
   There is no mode.
4. **D** Mean = $(42 + 89 + 92 + 85 + 89 + 74 + 63) / 7 = 76.1$
   Median = Put the numbers in numerical order then find the middle number: 42, 63, 74, 85, 89, 89, 92. 85 is the median.
   Mode = 89 is the number that happens the most.

   Since the mean and the median are both numbers in the middle of Abib's test scores, either one of these would be the best measure of central tendency.

5. **C** Put all of the choices in numerical order. The middle number in the set of data has to be 51.

6. **B** Mean = (0 + 84 + 86 + 90 + 92 + 92) / 6 = 74. The median = (86 + 90) / 2 = 88. The mode is 92. Look for the number that has the majority of the data near it. The median is the best choice.

7. **C** Mean = 60, median = 59, and mode = 100. Choice C is the only true statement.

8. **D** Mean = 20, median = 19, and mode = 18. Choice D is the only false statement.

Short Response

9. **Part A.**

Find the mean, median, and mode of the data.

$$\text{Mean} = \frac{44 + 37 + 22 + 11 + 17 + 25 + 34 + 17}{8} = \frac{207}{8} = 25.875$$

Median = Order the numbers: 11, 17, 17, 22, 25, 34, 37, 44.
The middle number is between 22 and 25.
So (22 + 25) / 2 = 23.5.

Mode = The number that occurs the most is 17.

The measure of central tendency that would encourage people to come to the park is the mode.

Part B.

The mode is correct because it is the smallest wait time in line for rides.

CIRCLE GRAPHS

1. **D** Take 50% of 360°: .50(360) = 180°

2. **C** Take 30% of 360°: .30(360) = 108°

3. **B** Find out what percent 72° is out of 360°: $\frac{72}{360} = 0.2 \rightarrow 20\%$

4. **C** Find out what percent 126° is out of 360°: $\frac{126}{360} = 0.35 \rightarrow 35\%$

5. C Since fishing is 15% of the circle, take 15% of 360:
$.15(360) = 54°$

6. C Since theme parks is 35% of the circle, take 35% of 360:
$.35(360) = 126°$

7. D Six people out of 20 own a dog. Change this to a percent:
$\frac{6}{20} = 0.3$

Take 30% of 360 to find what the central angle would be in the circle:
$.30(360) = 108°$

8. B Find what percent each pet is out of 360°:
Cats: $\frac{8}{20} = \frac{96}{360} = 96°$

Short Response

9. Part A.

To make the circle graph:

- Write a ratio of the number of students per item out of the total number of students.
- Change each ratio to a percent and multiply this by 360 to find the number of degrees.
- Using your protractor, make a circle graph.

Item	Number of Students	Ratio	Decimal (360)	Degrees
iPod	6	6/30	.2 (360)	72°
Cell phone	12	12/30	.4 (360)	144°
PlayStation	3	3/20	.1 (360)	36°
Computer	9	9/30	.3 (360)	108°

GIFTS STUDENTS WANT

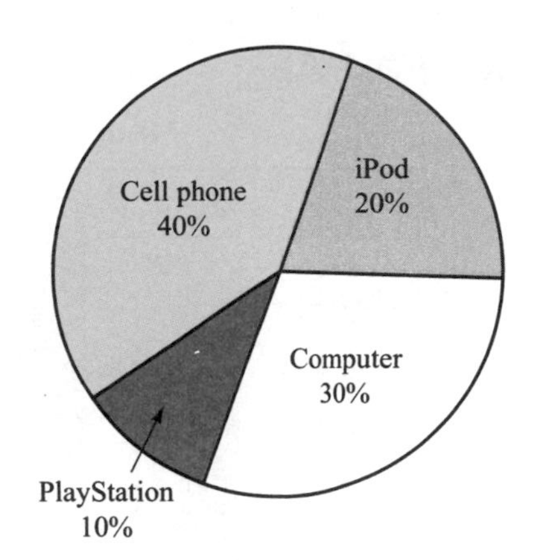

Part B. 20% of the students want an iPod.

DOUBLE BAR GRAPHS

1. **A** Look at both bars for Tuesday. Sam spent 30 minutes on the Internet. John spent 20 minutes. 30 – 20 = 10 minutes difference.
2. **B** Add the heights of each bar for John: 20 + 20 + 40 + 10 + 50 = 140
   Add the heights of each bar for Sam: 10 + 30 + 15 + 10 + 20 = 85
   Subtract Sam's total from John's total: 140 – 85 = 55
3. **C** On Thursday the height of the bars is the same.
4. **B** Girls = 10 + 40 + 20 = 70. Boys = 20 + 30 + 30 = 80.
   80 – 70 = 10.
5. **B** 40 – 10 = 30.
6. **C** Seventh-grade girls = 40 and eighth-grade girls = 20.
   40 – 20 = 20
7. **B** 80 – 60 = 20
8. **D** 90 – 60 = 30
9. **B** 2003: 60 + 60 = 120
   2007: 80 + 40 = 120

Short Response

10. Part A.

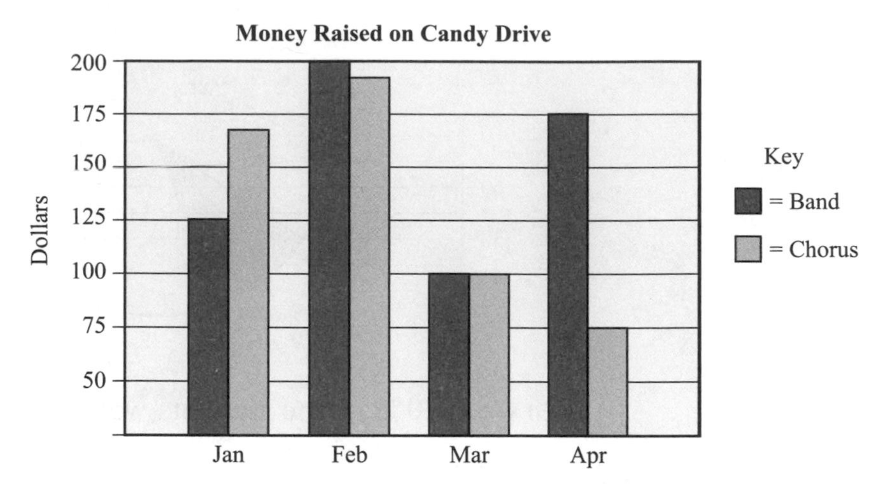

Part B. Add the months for band and chorus. The band sold \$600 in candy and the chorus sold \$535. Therefore, the band sold more candy.

DOUBLE LINE GRAPHS

1. **C** Plant 1 is represented by the solid line. The height on Thursday is 3 inches.
2. **A** Look at the points for both lines on Friday. Plant 2 (dotted) is at 4 inches. Plant 1 (solid) is at 3 inches. Plant 2 is 1 inch taller than Plant 1.
3. **D** Choice D is the only true statement.
4. **C** Between 2005 and 2006 there is a difference of 20,000.
5. **C** 50,000 – 20,000 = 30,000
6. **B** Choice B is the only true statement.

Short Response

7. **Part A.**

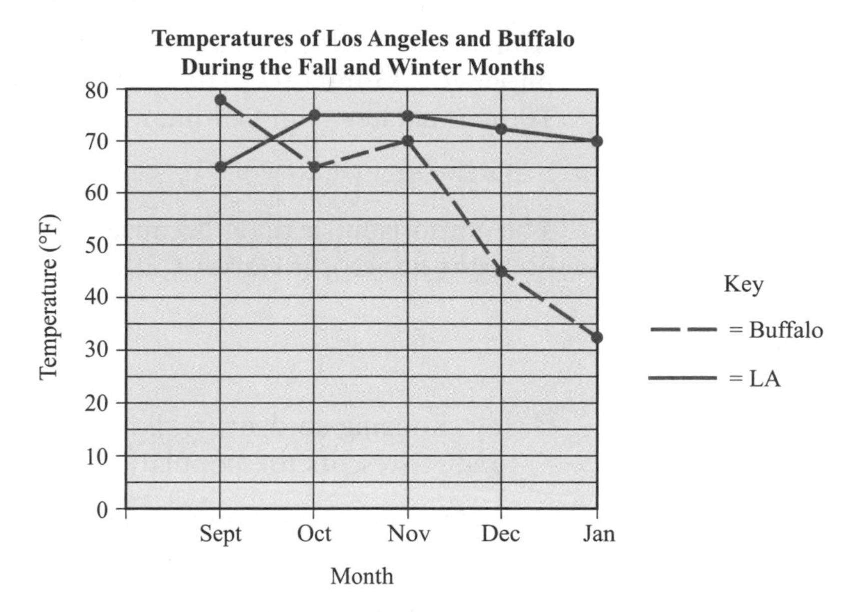

Part B. January shows the largest gap in temperature—January went from 65° to 32°.

MISLEADING STATISTICS

1. **C** Since the scale started at zero and went to seven, it should be 0, 7, 14, etc., not 0, 7, 8, 9, 10.
2. **C** "People do not order small subs" makes you think the number of small subs sold should equal zero, when in fact it equals three.
3. **D** The *y*-axis starts at 11 and it should start at zero.
4. **C** Eleven people like nachos and six people like hamburgers, so five more people like nachos than hamburgers is a true statement.
5. **A** The vertical scale is numbered 2, 4, 6, 7, 11, which is not evenly spaced.
6. **B** District C has the fewest number of snow days is the only true statement.
7. **C** At 10:00 A.M. the temperature was 60°F and at 12 P.M. it was 80°F. This is a 20° difference.

Short Response

8. **Part A.**

   Amber was on the phone for 25 minutes and Caitlin was on the phone for 35 minutes. Therefore, Amber talked on the phone 10 minutes less than Caitlin.

   Part B.

   The graph is misleading because the vertical axis (y-axis) does not start at zero. It starts at 20 and then counts by 5s.

SAMPLING

1. **B** By choosing students walking into the school, it is random and represents the population in the school.
2. **D** Calling every tenth person in the Wayne county phone book is random and surveys people in Wayne county.
3. **B** Surveying every other house in the neighborhood is random and would be a good representation of people in the neighborhood.
4. **B** Surveying people as they walk into the grocery store is random and will allow Drew to ask people of all ages who represent the people in his town.
5. **C** To get the best results, all 300 votes should have been counted in the tally.

Short Response

6. **Part A.** Yes.

   Part B. This is a good sampling technique because Skyler's sample represents the appropriate age group and she has a large sample.

Chapter 10

SOLUTIONS TO PRACTICE TEST 1

MULTIPLE CHOICE

1. **D** $2x - 3 = 17$
   $\quad +3 \quad +3$
   $\frac{2x}{2} = \frac{20}{2}$
   $x = 10$

2. **B** $502{,}000 \rightarrow 5.02 \times 10^5$

3. **B** A bar graph would the best graph because the cookies are different objects.

4. **D** 5km = _____m. Move 3 steps to the right on the metric staircase, so multiply by 1000. 5km = 5000 meters.

5. **C** The factors of 6 are 1, 2, 3, 6.
   The factors of 18 are 1, 2, 3, 6, 9, 18.
   The factors of 36 are 1, 2, 3, 4, 6, 9, 12, 18, 36.
   The highest number 6, 18, and 36 have in common is 6.

6. **A** $2^3 \cdot 2^2 = 2 \cdot 2 \cdot 2 \cdot 2 \cdot 2 = 2^5$

7. **C** The highest number is 136 and the lowest number is 16.
   The range is equal to 136 – 16 = 120.

8. **A** Substitute the values in for x, y, and w. Follow the order of operations.
   $xy - wx$
   $5(3) - 2(5)$
   $15 - 2(5)$
   $15 - 10$
   5

9. **B** $\frac{7x}{7} \geq \frac{91}{7}$
   $x \geq 13$

10. B Count the number of squares inside the figure.

11. D $7 - (-15)$
$7 + 15$
22

12. C Volume = lwh. V = 14(10)(6) = 840 cu. in.

13. D Remember that less than comes second in the problem.
Four times a number → $4n$
Two less than → -2
$4n - 2$

14. B One solution is to write out each number in standard form and look for the largest number.

15. A Area = 18. Since the sandbox is a square, all sides are equal.
Area = s^2
$18 = s^2$
$\sqrt{18} = s$

Since the problem says to estimate, the $\sqrt{18}$ is closest to the $\sqrt{16} = 4$.

16. D $\sqrt{25} = 5$. Five is not an irrational number because it ends.

17. C The angles in a quadrilateral add up to 360°. 120° + 45° + 85° = 250°.
360° − 250° = 110°.

18. D There are 22 students total. 11 students like math and science, which is half of the students. This is a true statement based on the table.

19. B Mean = (4 + 8 + 12 + 16 + 35 + 87) / 6 = 27
Median = (12 + 16) / 2 = 14
Mode = none
The median is a better choice because it has three numbers above it and three numbers below it. The mean only has two numbers above it and four numbers below it.

20. C Find multiples of 4, 6, and 18.

4 4, 8, 12, 16, 20, 24, 28, 32, 36, 40
6 6, 12, 18, 24, 30, 36
18 18, 36

36 is the smallest number that appears as a multiple for each number.

21. B $8^2 = 64$ and $9^2 = 81$, so $\sqrt{79}$ is between 8 and 9.

22. B Move 4 units to the left and 3 units down. (–4, –3)

23. B The metric unit is kilometers. None of the other choices are metric.

24. B The figure is a cylinder because it has two faces that are circles.

25. D 5 kiloliters = _____ liters
1 kiloliter = 1000 liters, so multiply by 5 to get kiloliters.
5(1000) = 5000.

26. B 36.04 rounds to 36. 5.96 rounds to 6. 36 ÷ 6 = 6.

27. C Set up a proportion. $\frac{\text{ounces}}{\text{loads}}$

$\frac{10}{6} = \frac{25}{x}$
$\frac{10x}{10} = \frac{150}{10}$
$x = 15$

28. B There are 4 outcomes on the spinner; A, B, C, D and 2 outcomes for the coin; H or T. So 4(2) = 8 possible outcomes.

29. D A bag of grapes is food. A produce scale measures the mass of food.

30. D The code has to be 3 digits. _____ _____ _____
There are 9 choices for each spot. So 9(9)(9) = 729 possible choices.

SHORT RESPONSE

31. Part A.

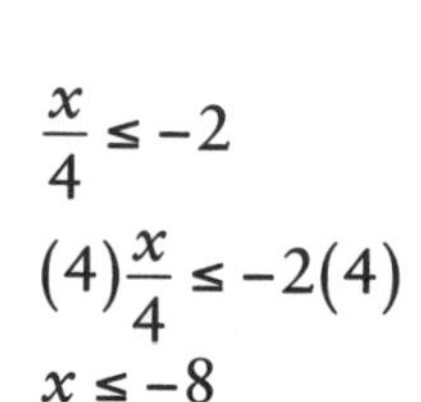

Part B.

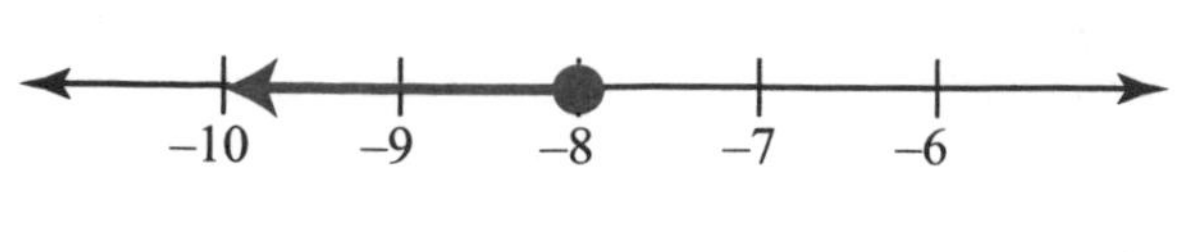

32. Part A. $2 \times 2 \times 2 \times 3 \times 3 \times 7$

This is one possible factor tree. The factor tree can start with any two factors of 820. You know that your tree is correct if your final answer is the same.

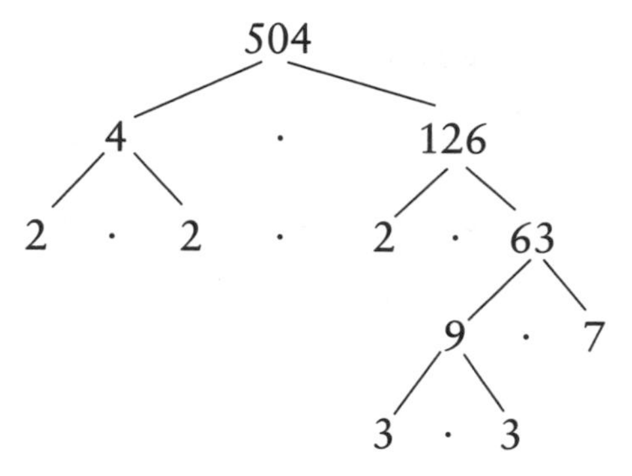

Part B. $2^3 \times 3^2 \times 7$

33. Part A. $3a - 2$

Part B. 34 years old

$3a - 2$
$3(12) - 2$
$36 - 2$
34

Part C. 6 years old

$3a - 2 = 16$
$\quad +2 \quad +2$
$\frac{3a}{3} = \frac{18}{3}$
$a = 6$

34. To estimate the surface area, first round the decimals.

height 6.2 → 6 length 14.3 → 14 width 3.8 → 4

SA = $2wl + 2lh + 2wh$
SA = $2(4)(14) + 2(14)(6) + 2(4)(6)$
SA = $112 + 168 + 48$
SA = 328 sq. in.

35. The distance around the merry-go-round is 27 feet. This is the circumference. Use the circumference formula to solve for the diameter. Since a value for π was not given, use the π button on your calculator when computing the answer.

$C = \pi d$
$\frac{27}{\pi} = \frac{\pi d}{\pi}$
$8.594366 = d$
$8.6 \text{ ft} = d$

36. Part A. Completed double line graph

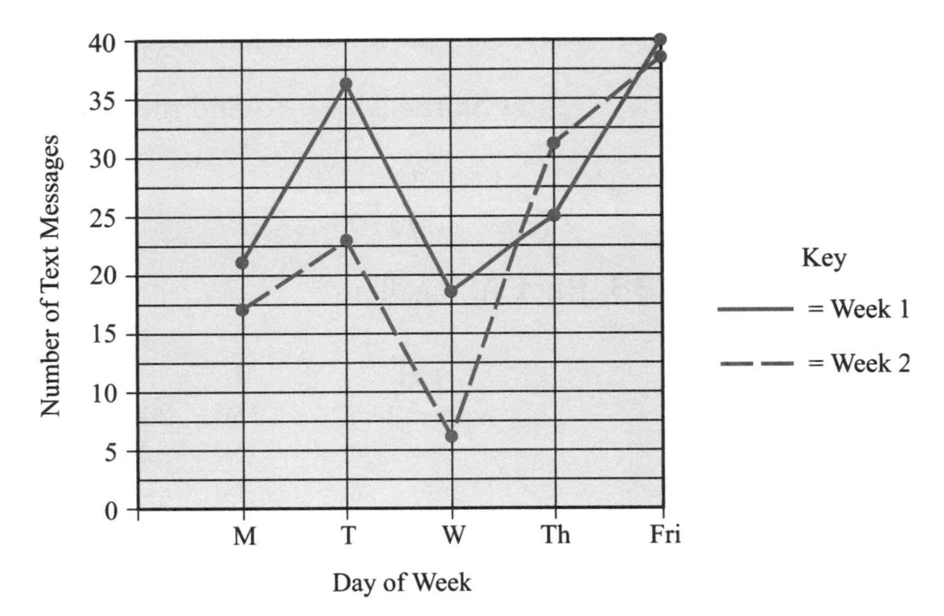

Part B. There are many possible conclusions. The statement has to be a true fact based on the graph. One possible conclusion: When comparing the two weeks, the only day Breann did not text message more in Week 1 than Week 2 was on Thursday.

Part C. Five text messages
The average number of text messages for Week 1 is $(22 + 36 + 18 + 25 + 41) / 5 = 28.4$. The average number of text messages for Week 2 is $(17 + 23 + 6 + 32 + 39) / 5 = 23.4$. The difference is $28.4 - 23.4 = 5$.

37. Part A.

$$\frac{2}{25} \cdot \frac{1}{24} = \frac{1}{300}$$

Part B.

There are 25 Skittles in the bag. The probability of getting a yellow Skittle on the first draw is $\frac{2}{25}$. Since Tommy does not return the Skittle to the bag, there are now 24 Skittles remaining. Since the first Skittle he picked was yellow, there is only one yellow Skittle left. The probability of getting a second yellow Skittle is $\frac{1}{24}$. To find the answer you multiply. So $\frac{2}{25} \cdot \frac{1}{24} = \frac{1}{300}$.

38. Part A. 7 gallons

$$\frac{\text{gal}}{\text{square feet}} \qquad \frac{1}{350} = \frac{x}{2450}$$

$$\frac{350x}{350} = \frac{2450}{35}$$

$$x = 7 \text{ gallons}$$

Part B. 28 quarts

There are 4 quarts in a gallon, so 7(4) = 28 quarts.

Chapter 11

SOLUTIONS TO PRACTICE TEST 2

MULTIPLE CHOICE

1. **B** Make a factor tree.

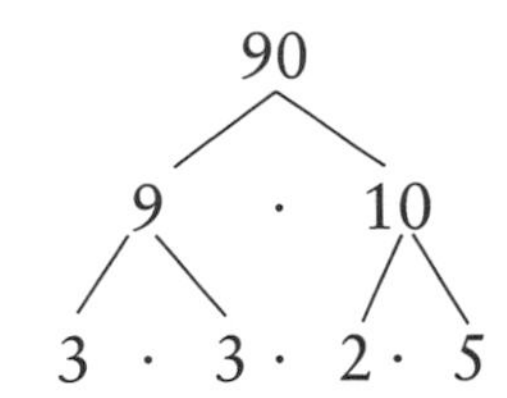

$3 \cdot 3 \cdot 2 \cdot 5 = 2 \cdot 3^2 \cdot 5$

2. **C** 2 gal = _____ pts.

Move 2 steps to the right on the customary staircase.
Multiply by 4 and 2.
So $2(4)(2) = 16$.

3. **A** The factors of 56 are 1, 2, 4, 7, 8, 14, 28, 56. The first four factors are 1, 2, 4, 7.

4. **C** The rectangular prism has six faces; front, back, top, bottom, and two sides.

5. **B** Whole numbers = {0, 1, 2, 3, 4,...}. The number 3 is a whole number.

6. **C** $2.4 \times 10^3 \rightarrow 2400$

7. **B** $|-20| + 15 \div 5 \cdot 2^2$
$20 + 15 \div 5 \cdot 4$
$20 + 3 \cdot 4$
$20 + 12$
32

8. C $\frac{x}{4} - 3 = 8$

$\underline{\quad +3 \;\; +3}$

$(4)\frac{x}{4} = 11(4)$

$x = 44$

9. B $2^6 \div 2^3 = 64 \div 8 = 8$

10. A $\frac{2}{5} \cdot \frac{2}{5} = \frac{4}{25}$

11. A 10 – 15 + 25

10 + (–15) + 25

–5 + 25

20

12. D $\frac{2}{9} = \frac{5}{x}$

$\frac{2x}{2} = \frac{45}{2}$

$x = \$22.50$

13. D This is a true statement because there are 2 students in overlap between only breakfast and dinner.

14. B $\frac{0.24}{12} = 0.02$

15. D The sum of three times x and eleven → $3x + 11$

is equal to fourteen → = 14

$3x + 11 = 1\ 4$

16. C There are 60 total students. 15 out of 60 are in the newspaper club. $\frac{15}{60} = \frac{1}{4}$

17. A Natural numbers = {1, 2, 3,...} and whole numbers = {0, 1, 2, 3,...}. Three is in both of these sets.

18. C $10 > x - 8$

$\underline{+8 \quad +8}$

$18 > x$ This reads as $x < 18$ or x is less than 18.

19. C $V = \pi r^2 h$
$V = \pi(4^2)(14)$
$V = \pi(16)(14)$
$V = 224\pi$

20. C The central angle must pass through the center *C*. $\angle BCE$ is the central angle.

21. C There is a $\frac{2}{5}$ chance of getting an even number. Set this equal to $\frac{x}{160}$ to see how many times out of 160 Damon will get an even number.

$\frac{2}{5} = \frac{x}{160}$
$\frac{5x}{5} = \frac{320}{5}$
$x = 64$

22. A $10^{-3} = \frac{1}{10^3} = \frac{1}{1000} = 0.001$

23. B Find the LCM of 7 and 12. 84 is the least common multiple.
7   7, 14, 21, 28, 35, 42, 49, 56, 63, 70, 77, 84
12   12, 24, 36, 48, 60, 72, 84

24. D The highest number is 98, the lowest number is 21. 98 – 21 = 77.

25. D In Game 4 Katherine scored 2 points, in Game 1 she scored 6 points. This is three times as many points.

26. C $\frac{9}{20}$ students said Friday was their favorite day of the week.

Set this equal to $\frac{x}{100}$ to make it a percent.

$\frac{9}{20} = \frac{x}{100}$
$\frac{20x}{20} = \frac{900}{20}$
$x = 45\%$

27. B
$$V = lwh$$
$$108 = 9(3)(h)$$
$$\frac{108}{27} = \frac{27h}{27}$$
$$h = 4$$

28. C
$$C = \pi d$$
$$\frac{43.96}{3.14} = \frac{3.14d}{3.14}$$
$$d = 14$$

29. B
$4m + 3n - 5s + 3$
$4(6) + 3(2) - 5(3) + 3$
$24 + 6 - 15 + 3$
$30 - 15 + 3$
$15 + 3$
18

30. A To get a sample that would survey the different types of students in the middle school, the best selection would be to survey students in sixth, seventh, and eighth grades during their lunch periods.

SHORT RESPONSE

31. Part A. $\sqrt{7}$, $\frac{13}{2}$, 10.5, $\sqrt{121}$, 3^3, 4.3×10^2, 4300

Change each number to as decimal.

4.3×10^2	3^3	10.5	$\frac{13}{2}$	$\sqrt{7}$	$\sqrt{121}$	4300
430	27	10.5	6.5	2.64...	11	4300
#6	#5	#3	#2	#1	#4	#7

Part B.
$\sqrt{121}$ is a rational number. The $\sqrt{121} = 11$. Since 11 ends and it can be written as a fraction, it is rational. An irrational number is a decimal that does not repeat and goes on forever.

32. Part A. $0.15m + 25 = 100$

Part B. 500 minutes

$$\begin{aligned} 0.15m + 25 &= 100 \\ -25 \quad &\ \ -25 \\ \hline \frac{0.15m}{0.15} &= \frac{75}{0.15} \\ m &= 500 \end{aligned}$$

33. Part A.

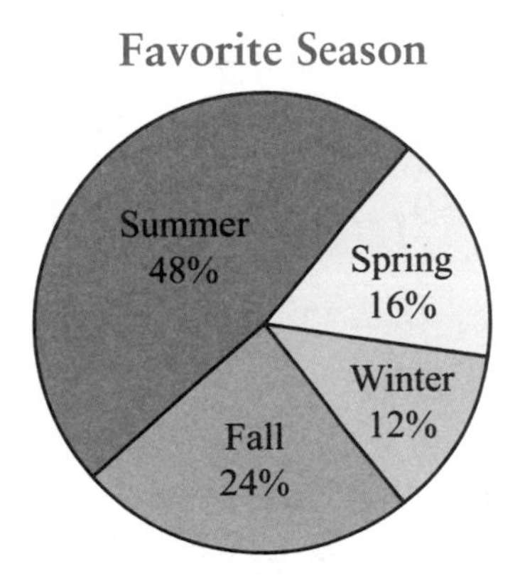

Part B. 48%. Set up a proportion to find what percent this is out of 100.

$$\frac{12}{25} = \frac{x}{100}$$
$$\frac{25x}{25} = \frac{1200}{25}$$
$$x = 48\%$$

34. 3 inches

$$\text{A} = \pi r^2$$
$$\frac{28.26}{3.14} = \frac{3.14r^2}{3.14}$$
$$9 = r^2$$
$$3 = r$$

35. Part A. 77°F

$\frac{9}{5}C + 32$

$\frac{9}{5}(25) + 32$

$45 + 32$

77°F

Part B. 14°F

Convert 20°C to °F and then find the difference between the two temperatures.

$\frac{9}{5}C + 32$ 82°F – 68°F = 14°F

$\frac{9}{5}(20) + 32$

$36 + 32$

68°F

36. $SA = 2\pi rh + 2\pi r^2$
$SA = 2(3.14)(6)(12) + 2(3.14)(6^2)$
$SA = 452.16 + 226.08$
$SA = 678.24$

37. Part A. 128 feet

To find one side of the square yard, take the square root of 1024.
$\sqrt{1024} = 32$ feet. To find the perimeter of the yard, multiply 32 by 4 since there are four sides on a square.
32(4) = 128 feet

Part B. 13 lengths

To estimate, round 128 to 130 feet. Since fencing is sold in 10-foot lengths, see how many times 10 will go into 130.
$\frac{130}{10} = 13$ feet

38. Part A.

Mean = (80 + 100 + 60 + 100 + 70) / 5 = 82
Median = First put the numbers in numerical order: 60, 70, 80, 100, 100. Then find the middle number. The median is 80.
Mode = 100, the number that occurs the most.
Range = H – L
100 – 60 = 40

Part B.

The mode (100) is the best measure of central tendency to use because it is the highest number and therefore is the best grade.

Part C.

Scott's grade of 100 is misleading because it might seem that if his grade is 100, he has gotten 100s on all of his tests, and that is not true. The mean or the median is a more accurate grade. The grades of 80 or 82 are in the middle of all of his scores, which range from 60 to 100.